Introduction to
Soil Microbiology

Introduction to Soil Microbiology

Abdiel Edwards
Editor

KOROS PRESS LIMITED
London, UK

Introduction to Soil Microbiology

© 2012
Printed in 2017 for Sale in the Indian Subcontinent

Published by
Koros Press Limited
3 The Pines, Rubery B45 9FF, Rednal,
Birmingham, United Kingdom

Tel.: +44-7826-930152
Email: info@korospress.com
www.korospress.com

ISBN: 978-1-78163-104-1

Editor: Abdiel Edwards

Printed in UK

British Library Cataloguing in Publication Data
A CIP record for this book is available from the British Library

10 9 8 7 6 5 4 3 2 1

No part of this publication may be reproduced, stored in a retrieval system or transmitted in any form or by any means, electronic, mechanical, photocopying, recording, scanning or otherwise without prior written permission of the publisher.

Reasonable efforts have been made to publish reliable data and information, but the authors, editors, and the publisher cannot assume responsibility for the legality of all materials or the consequences of their use. The authors, editors, and the publisher have attempted to trace the copyright holders of all materials in this publication and express regret to copyright holders if permission to publish has not been obtained. If any copyright material has not been acknowledged, let us know so we may rectify in any future reprint.

Exclusively distributed by CBS Publishers & Distributors Pvt. Ltd.

Sales & Distribution Rights only for India, Pakistan, Bangladesh, Sri Lanka, Nepal and Bhutan.This book is not to be sold outside these territories.

Contents

Preface   *vii*

1. **Introduction**   1
   Soil • Soil Horizons • USDA Soil Taxonomy • Soil Solutions • Soil-forming Materials • Soil Texture • Principal Pedogenic Processes • Soil Classification • Canadian System of Soil Classification • Soils of India • Different Types of Soil • Humus Soil • Soil Pollution Causes and Effects • Soil Moisture Levels • Soil Moisture Metres • Soil Moisture Sensors • Soil Electrical Conductivity Variability • Methods • Results and Discussion

2. **Soil Testing : How to Test Soil pH**   47
   Soil pH Testing With Test Kits • Lowering Soil pH - How to Lower Soil pH Level • Soil Types • Acidic Soil • Soil Horizon Layers • Role of the Extension Service in Soil Testing • Soil Analysis: A key to Soil Nutrient Mangement • Interpretation of Soil pH • Soil pH Levels • Soil pH Test • Soil Pollution Facts • Analysis and Collection of Soil Samples • Tree Roots Effects on Soil • Garden Soil Preparation • Potting Soil Recipe

3. **Soil Amendments**   85
   Types of Soil Amendments • Land Pollution • Facts about Land Pollution • 10 Ways to Conserve Soil • Wind Erosion and Deposition • Soil Erosion Solutions • Soil Erosion Causes and Solutions • Soil Erosion Facts • What is Soil Erosion • Soil Erosion Control • Soil Erosion Control Methods • Wind Erosion • Soil Erosion Prevention • Soil Conservation Methods

4. **Soil Conservation**   114
   Ways to Conserve Soil • Soil Erosion Causes • Natural Causes of Soil Erosion • Erosion Control Methods • How are Rivers Formed? • How to Improve Clay Soil • Growing Seeds Without Soil • Soil Testing • Granite Rock Facts • Acidic Soil Plants • Types of Landforms • Rock: Types of Rocks • Different Rock Types • Moisture Metre: Moisture Detection and Analysis

5. **Soil Profile** 144
   Organic Matter • Soil Organic Matter • Organic Matter in Water • Soil Organic Matter • Organic Matter in Virgin and Cultivated Soils • Managing Soil Organic Matter • Biomass • Detritus

6. **Organic Geochemistry** 170
   Total Organic Carbon • Walkely-Black (1934) Modified Method • Biotic Material • Organic Farming • Soil Management • Loss of Soil Organic Matter and its Restoration • Fuel for the Plant-Food Production Factory • Problem of Maintaining a Liberal Supply of Soil Organic Matter • Interrelation of Soil Organic Matter with Nitrogen and Minerals • Soil Salinisation • Soil Susceptibility to Compaction • Landslides in Europe • Landslides and the EU Soil Thematic Strategy

7. **Soil Sealing** 228
   Soil Contamination • Soil Biodiversity • The Value of Soil Biodiversity • Soil Sampling • Organic Carbon Stock in Mineral Soils of European Union • Soil Organic Matter, Green Manures and Cover Crops for Nematode Management • Manure Management and Effects of Manure on the Environment • Soil Texture • Soil Salinity and Plant Sensitivity to Salts • Organic Farming : Manures

   Bibliography 259

   Index 263

Preface

It is the outer, loose material of earth's surface which is distinctly different from the underlying bedrock and the region which support plant life. Agriculturally, soil is the region which supports the plant life by providing mechanical support and nutrients required for growth. From the microbiologist view point, soil is one of the most dynamic sites of biological interactions in the nature. It is the region where most of the physical, biological and biochemical reactions related to decomposition of organic weathering of parent rock take place. Living organisms both plant and animal types constitute an important component of soil. Though these organisms form only afraction (less than one percent) of the total soil mass, but they play important role in supporting plant communities on the earth surface. While studying the scope and importance of soil microbiology, soil-plant-animal ecosystem as such must be taken into account.

Bacteria are the smallest organisms in the soil and are the only soil microorganisms that are prokaryotic. All of the other microorganisms are eukaryotic, which means they have a more advanced cell structure with internal organelles and the advanced ability to reproduce sexually. A prokaryote has a very simple cell structure with no internal organelles. Bacteria are the most abundant microorganisms in the soil, and serve many important purposes, one of those being nitrogen fixation among other biochemical processes. One of the most distinguished features of bacteria as a whole is their biochemical versatility. A species called *Pseudomonas* can metabolize a wide range of chemicals and fertilizers. In contrast, another species known as *Nitrobacters* can only derive its energy by turning nitrite into nitrate, which results in a gain of oxygen and is known also as oxidation. Furthermore, the species *Clostridium* is also an example of bacteria's versatility because it, unlike most species, can actually grow in the absence of oxygen. Bacteria are responsible for the process of nitrogen fixation, which is the conversion of atmospheric nitrogen into nitrogen which can be used by plants to uptake. Autotrophic bacteria, or bacteria that derives its energy making its own food by oxidation, like the *Nitrobacters* species, rather than feeding on plants or other organisms. The bacteria that are autotrophic are responsible for nitrogen fixation, and the amount of autotrophic bacteria is small compared to heterotrophic bacteria (the opposite of autotrophic bacteria, heterotrophic bacteria acquires energy by consuming plants or other microorganisms), but are very important because almost every plant and organism require nitrogen in some way, and would have no way of obtaining it if not for nitrogen-fixing bacteria. Actinomycetes are soil microorganisms. They are a type of bacteria. They are similar to both bacteria and fungi, and have characteristics linking them to both groups. Actinomycetes are often believed to be the missing evolutionary link between bacteria and fungi, but they have many more characteristics in common with bacteria than they do fungi. Actinomycetes are similar to bacteria because they, like bacteria, are prokaryotic, are sensitive to antibacterial and affected in the same way that bacteria is by them. Actinomycetes can hardly be distinguished from bacteria at its early stages because of how much they resemble bacteria in size, shape and gram-staining properties. Gram staining is a common technique used to classify organisms into two main groups: Gram-positive and Gram-negative, by staining organisms to distinguish its cell wall properties. Gram-positive means that the cell has a thin, penetrable cell wall and gram-negative means the opposite, that the cell wall is thick and difficult to penetrate. Cell wall properties can help distinguish different types of microorganisms from each other.

The book keeps the learners abreast with the current trends and concepts in soil microbiology and soil biotechnology.

—Editor

Chapter 1
Introduction

Soil

Soil is a natural body consisting of layers (soil horizons) of mineral constituents of variable thicknesses, which differ from the parent materials in their morphological, physical, chemical, and mineralogical characteristics. Strictly speaking, soil is the depth of regolith that influence and have been influenced by plant roots.

Figure: Darkened topsoil and reddish subsoil layers are typical in some regions.

Soil is composed of particles of broken rock that have been altered by chemical and mechanical processes that include weathering and erosion. Soil differs from its parent rock due to interactions between the lithosphere, hydrosphere, atmosphere, and the biosphere. It is a mixture of mineral and organic constituents that are in solid, gaseous and aqueous states. Soil is commonly referred to as earth or dirt.

Soil forms a structure that is filled with pore spaces, and can be thought of as a mixture of solids, water and air (gas). Accordingly, soils are often treated as a three state system. Most soils have a density between 1 and 2 g/cm^3. Little of the soil composition of planet Earth is older than the Tertiary and most no older than the Pleistocene. In engineering, soil is referred to as regolith, or loose rock material.

History of the Study of Soil

The history of the study of soil is intimately tied to our urgent need to provide food for ourselves and forage for our animals.

Columella's Husbandry, circa 60 A.D. was used by 15 generations (450 years) of those encompassed by the Roman Empire until its collapse. From the fall of Rome to the French Revolution, knowledge of soil and agriculture was passed on from parent to child and as a result, crop yields were low. During the Dark Ages for Europe, Yahya Ibn_al-'Awwam's handbook guided the people of North Africa, Spain and the Middle East with its emphasis on irrigation, a translation of which was finally carried to the southwest of the United States.

Jethro Tull, an English gentleman, introduced in 1701 an improved grain drill that systemised the planting of seed and invented a horse-drawn weed hoe, the two of which allowed fields, once choked with weeds to be brought back to production and seed to be used more economically. Tull however introduced the mistaken idea that manure introduced weed seeds, and that fields should be plowed in order to pulverize the soil and so release the locked up nutrients. His ideas were taken up and carried to their extremes in the 20th century, where farmers repeatedly plowed fields far beyond what was necessary to control weeds, resulting in the dust bowl of the panhandle areas of Texas and Oklahoma of the United States.

The two course system of a year of wheat followed by a year of fallow was replaced in the 18th century by the Norfolk four-course system wherein wheat was grown in the first year, turnips the second, followed by barley, with clover and ryegrass together, in the third. The taller barley was harvested in the third year while the clover and ryegrass were grazed or cut for feed in the fourth. The turnips fed cattle and sheep in the winter. The fodder crops produced large supplies of animal manure which returned nutrients to the soil.

Experiments into what made plants grow first lead to the idea that the ash left behind when plant matter was burnt was the essential element, overlooked the role of nitrogen which is not left on the ground after combustion. Jan Baptista van Helmont thought he had proved water to be the essential element from his famous experiment with a willow tree grown in a carefully controlled conditions in which only water was added and after five years of growth was removed and weighed, roots and all and found to weigh 165 pounds The oven dried soil, originally 200 pounds was again dried and weighed and found to have lost only two ounces which van Helmont reasonably explained as experimental error and assumed that the soil had in fact lost nothing. As rain water was the only thing added by the experimenter he concluded that water was the essential element in plant life. In fact the two ounces lost from the soil were the minerals taken up by the willow tree during its growth.

John Woodward experimented with various types of water ranging from clean to muddy and found muddy water the best and so he concluded earthy matter was the essential element. Others concluded it was humus in the soil that passed some essence to the growing plant.

The French chemist Antonine Lavoisier showed that plants and animals must "combust" oxygen internally to live and was able to deduce that most of the 165 pound weight of Van Helmont's willow tree derived from air. The chemical basis of nutrients delivered to the soil in manure was emphasized and in mid 19th century chemical fertilizers were used but the dynamic interaction of soil and its life forms awaited discovery.

It was known that nitrogen was essential for growth and in 1880 the presence of Rhizobium bacteria in the roots of legumes explained the increase of nitrogen in soils so cultivated. The importance of life forms in soil were finally recognised.

Crop rotation, mechanization, chemical and natural fertilizers lead to a doubling of wheat yields in western Europe between 1800 to 1900.

Soil Forming Factors

Soil formation, or pedogenesis, is the combined effect of physical, chemical, biological, and anthropogenic processes on soil parent material. Soil genesis involves processes that develop layers or horizons in the soil profile. These processes involve additions, losses, transformations and translocations of material that compose the soil.

Minerals derived from weathered rocks undergo changes that cause the formation of secondary minerals and other compounds that are variably soluble in water. These constituents are moved (translocated) from one area of the soil to other areas by water and animal activity. The alteration and movement of materials within soil causes the formation of distinctive soil horizons.

The weathering of bedrock produces the purely mineral based parent material from which soils form. An example of soil development from bare rock occurs on recent lava flows in warm regions under heavy and very frequent rainfall. In such climates, plants become established very quickly on basaltic lava, even though there is very little organic material. The plants are supported by the porous rock as it is filled with nutrient-bearing water which carries dissolved minerals from rocks and guano. The developing plant roots, themselves are associated with mycorrhizal fungi that gradually break up the porous lava, and by these means organic matter and a finer mineral soil soon accumulates.

But even before it does, the predominantly porous broken lava in which the plant roots grow can be considered a soil. How the soil "life" cycle proceeds is influenced by at least five classic soil forming factors that are dynamically intertwined in shaping the way soil is developed, they include: parent material, regional climate, topography, biotic potential and the passage of time.

Parent Material

The material from which soil forms is called parent material. It includes: weathered primary bedrock; secondary material transported from other locations, e.g. colluvium and alluvium; deposits that are already present but mixed or altered in other ways - old soil formations, organic material; and anthropogenic materials, such as landfill or mine waste.

Soils that develop from their underlying parent rocks are called "residual soils", and have the same general chemistry as their parent rocks. Few soils form in such a manner.

Most soils derive from transported parent materials that have been moved by wind, water and gravity many miles.. Windblown material called loess, common in the Midwest of North America and in Central Asia, may have been moved many hundreds of miles. Cumulose parent material include peats and mucks, may develop in place from plant residues have been preserved by the low oxygen content of a high water table.

Weathering is the first stage in the transforming of parent material into soil material. In soils forming from bedrock, a thick layer of weathered material called saprolite may form. Saprolite is the result of weathering processes that include: hydrolysis (the division of a mineral into acid and base pairs by the splitting of intervening water molecules), chelation from organic compounds, hydration (the solution of minerals in water with resulting cation, anion pairs), and physical processes that include freezing and thawing. The mineralogical and chemical composition of the primary bedrock material, its physical features, including grain size and degree of consolidation, plus the rate and type of weathering, transforms the parent material into the different mineral components of soils.

Climate

Soil formation greatly depends on the climate, and soils show the distinctive characteristics of the climate zones in which they originate. Temperature and moisture affect weathering and leaching. Wind moves sand and smaller particles, especially in arid regions where there is little plant cover. The type and amount of precipitation influence soil formation by affecting the movement of ions and particles through the soil, and aid in the development of different soil profiles. The effectiveness of water in weathering parent rock material depends on seasonal and daily temperature fluctuations. The cycles of freezing and thawing is an effective mechanism that breaks up rocks and other consolidated materials. Temperature and precipitation rates affect vegetation cover, biological activity, and the rates of chemical reactions in the soil.

Biological Factors

Plants, animals, fungi, bacteria and humans affect soil formation. Animals and micro-organisms mix soils as they form burrows and pores allowing moisture and gases to move about. In the same way, plant roots open channels in soils. Plants with deep taproots can penetrate many metres through the different soil layers to bring up nutrients from deeper in the profile. Plants with fibrous roots that spread out near the soil surface, have roots that are easily decomposed, adding organic matter. Micro-organisms, including fungi and bacteria, affect chemical exchanges between roots and soil and act as a reserve of nutrients. Humans can impact soil formation by removing vegetation cover with erosion the result. They can also mix the different soil layers, restarting the soil formation process as less-weathered material is mixed with the more developed upper layers. Some soils may contain up to one million species of microbes per gram, most of those species being unknown, making soil the most abundant ecosystem on Earth.

Vegetation impacts soils in numerous ways. It can prevent erosion caused by excessive rain and the resulting surface runoff. Plants shade soils, keeping them cooler and slowing evaporation of soil moisture, or plants by way of transpiration can cause soils to lose moisture. Plants can form new chemicals which can break down or build up soil particles. The type and amount of vegetation depends on climate, land form topography, soil characteristics, and biological factors. Soil factors such as density, depth, chemistry, pH, temperature and moisture greatly affect the type of plants that can grow in a given location. Dead plants and dropped leaves and stems fall to the surface of the soil and decompose. There, organisms feed on them and mix the organic material with the upper soil layers; these added organic compounds become part of the soil formation process.

Time

Time is a factor in the interactions of all the above. Over time, soils evolve features dependent on the other forming factors. Soil formation is a time-responsive process that is dependent on how the other factors interplay with each other. Soil is always changing. It takes about 800 to 1000 years for a 2.5 cm thick layer of fertile soil to be formed in nature. For example, recently-deposited material from a flood exhibits no soil development because there has not been enough time for soil-forming activities. The original soil surface is buried, and the formation process must begin anew for this deposit. The long periods over which change occurs and its multiple influences mean that simple soils are rare, resulting in the formation of soil horizons. While soil can achieve relative stability of its properties for extended periods, the soil life cycle ultimately ends in soil conditions that leave it vulnerable to erosion. Despite the inevitability of soil retrogression and degradation, most soil cycles are long and productive.

Soil-forming factors continue to affect soils during their existence, even on "stable" landscapes that are long-enduring, some for millions of years. Materials are deposited on top and materials are blown or washed from the surface. With additions, removals and alterations, soils are always subject to new conditions. Whether these are slow or rapid changes depend on climate, landscape position and biological activity.

Characteristics

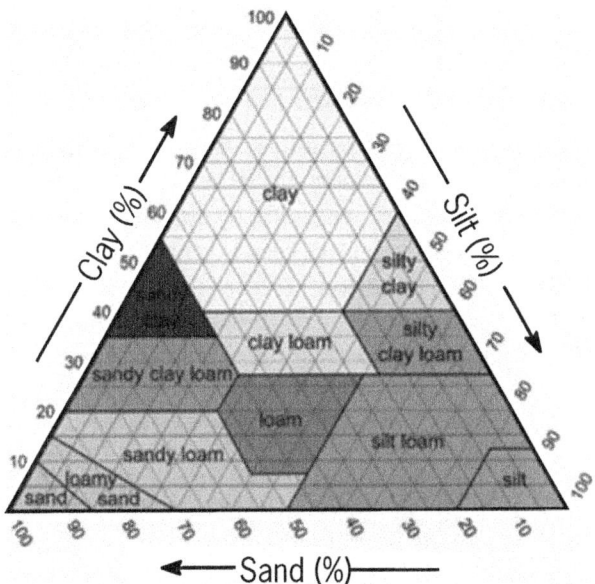

Figure: Soil types by clay, silt and sand composition as used by the USDA:

Figure: Iron rich soil near Paint Pots in Kootenay National Park of Canada.

On a volume basis a good quality soil is one that is 45% minerals, 25% water, 25% air, and 5% organic material, both live and dead. The mineral and organic components are considered a constant with the percentages of water and air the only variable parameters where the increase in one is balanced by the reduction in the other. The mineral components of soil may consist of a mixture of clay, sand, and silt. In the illustrated textural classification triangle the only soil that does not exhibit one of those predominately is called "loam." While even pure sand, silt or clay may be considered a soil, from the perspective of food production a loam soil with a small amount of organic material is considered ideal. The mineral constituents of a loam soil might be 40% sand, 40% silt and the balance 20% clay.

Soil colour is often the first impression one has when viewing soil. Striking colours and contrasting patterns are especially noticeable. The Red River (Mississippi watershed) carries sediment eroded from extensive reddish soils like Port Silt Loam in Oklahoma. The Yellow River in China carries yellow sediment from eroding loess soils. Mollisols in the Great Plains are darkened and enriched by organic matter. Podsols in boreal forests have highly contrasting layers due to acidity and leaching. Soil colour is primarily influenced by soil mineralogy. Many soil colours are due to various iron minerals.

The development and distribution of colour in a soil profile result from chemical and biological weathering, especially redox reactions. As the primary minerals in soil parent material weather, the elements combine into new and colourful compounds. Iron forms secondary minerals with a yellow or red colour, organic matter decomposes into black and brown compounds, and manganese, sulfur and nitrogen can form black mineral deposits. These pigments can produce various colour patterns within a soil. Aerobic conditions produce uniform or gradual colour changes, while reducing environments (an aerobic) result in disrupted colour flow with complex, mottled patterns and points of colour concentration.

Soil structure is the arrangement of soil particles into aggregates. These may have various shapes, sizes and degrees of development or expression. Soil structure affects aeration, water movement, resistance to erosion and plant root growth. Structure often gives clues to texture, organic matter content, biological activity, past soil evolution, human use, and chemical and mineralogical conditions under which the soil formed. If the soil is too high in clay, adding gypsum, washed river sand and organic matter will balance the composition. Adding organic matter to soil that is depleted in nutrients and too high in sand will boost the quality.

Soil texture refers to a soil's sand, silt and clay composition. Soil content affects soil behaviour, including the retention capacity for nutrients and water. Sand and silt are the products of physical weathering, while clay is the product of chemical weathering. Clay content has retention capacity for nutrients and water. Clay soils resist wind and water erosion better than silty and sandy soils, as the particles are bonded to each other. In medium-textured soils, clay is often moved downward through the soil profile and accumulates in the subsoil.

Soil resistivity is a measure of a soil's ability to retard the conduction of an electric current. The electrical resistivity of soil can affect the rate of galvanic corrosion of metallic structures in contact with the soil. Higher moisture content or increased electrolyte concentration can lower the resistivity and increase the conductivity thereby increasing the rate of corrosion. Soil resistivity values typically range from about 2 to 1000 Ù·m, but more extreme values are not unusual.

Soil Horizons

The naming of soil horizons is based on the type of material the horizons are composed of; these materials reflect the duration of the specific processes used in soil formation. They are labelled using a short hand notation of letters and numbers. They are described and classified by their colour, size, texture, structure, consistency, root quantity, pH, voids, boundary characteristics, and if they have nodules or concretions. Any one soil profile does not have all the major horizons covered below; soils may have few or many horizons.

The exposure of parent material to favourable conditions produces initial soils that are suitable for plant growth. Plant growth often results in the accumulation of organic residues, the accumulated organic layer is called the O horizon. Biological organisms colonise and break down organic materials, making available nutrients that other plants and animals can live on. After sufficient time a distinctive organic surface layer forms with humus which is called the A horizon.

Classification

Soil is classified into categories in order to understand relationships between different soils and to determine the usefulness of a soil for a particular use. One of the first classification systems was developed by the Russian scientist Dokuchaev around 1880. It was modified a number of times by American and European researchers, and developed into the system commonly used until the 1960s. It was based on the idea that soils have a particular morphology based on the materials and factors that form them. In the 1960s, a different classification system began to emerge, that focused on soil morphology instead of parental materials and soil-forming factors. Since then it has undergone further modifications. The World Reference Base for Soil Resources (WRB) aims to establish an international reference base for soil classification.

USDA Soil Taxonomy

In the United States, *soil orders* are the highest hierarchical level of soil classification in the USDA Soil Taxonomy classification system. Names of the orders end with the suffix *-sol*. There are 12 soil orders in Soil Taxonomy:

- Entisol - recently formed soils that lack well-developed horizons. Commonly found on unconsolidated sediments like sand, some have an A horizon on top of bedrock.
- Vertisol - inverted soils. They tend to swell when wet and shrink upon drying, often forming deep cracks that surface layers can fall into.
- Inceptisol - young soils. They have subsurface horizon formation but show little eluviation and illuviation.
- Aridisol - dry soils forming under desert conditions. They include nearly 20% of soils on Earth. Soil formation is slow, and accumulated organic matter is scarce. They may have subsurface zones (calcic horizons) where calcium carbonates have accumulated from percolating water. Many aridiso soils have well-developed Bt horizons showing clay movement from past periods of greater moisture.
- Mollisol - soft soils with very thick A horizons.
- Spodosol - soils produced by podsolization. They are typical soils of coniferous and deciduous forests in cooler climates.
- Alfisol - soils with aluminium and iron. They have horizons of clay accumulation, and form where there is enough moisture and warmth for at least three months of plant growth.
- Ultisol - soils that are heavily leached.
- Oxisol - soil with heavy oxide content.

- Histosol - organic soils.
- Andisols - volcanic soils, which tend to be high in glass content.
- Gelisols - permafrost soils.

Organic Matter

Most living things in soils, including plants, insects, bacteria and fungi, are dependent on organic matter for nutrients and energy. Soils often have varying degrees of organic compounds in different states of decomposition. Many soils, including desert and rocky-gravel soils, have no or little organic matter. Soils that are all organic matter, such as peat (histosols), are infertile.

Humus

Humus refers to organic matter that has decomposed to a point where it is resistant to further breakdown or alteration. Humic acids and fulvic acids are important constituents of humus and typically form from plant residues like foliage, stems and roots. After death, these plant residues begin to decay, starting the formation of humus. Humus formation involves changes within the soil and plant residue, there is a reduction of water soluble constituents including cellulose and hemicellulose; as the residues are deposited and break down, humin, lignin and lignin complexes accumulate within the soil; as microorganisms live and feed on the decaying plant matter, an increase in proteins occurs.

Lignin is resistant to breakdown and accumulates within the soil; it also chemically reacts with amino acids which add to its resistance to decomposition, including enzymatic decomposition by microbes. Fats and waxes from plant matter have some resistance to decomposition and persist in soils for a while. Clay soils often have higher organic contents that persist longer than soils without clay. Proteins normally decompose readily, but when bound to clay particles they become more resistant to decomposition. Clay particles also absorb enzymes that would break down proteins. The addition of organic matter to clay soils can render the organic matter and any added nutrients inaccessible to plants and microbes for many years, since they can bind strongly to the clay. High soil tannin (polyphenol) content from plants can cause nitrogen to be sequestered by proteins or cause nitrogen immobilisation, also making nitrogen unavailable to plants. Humus formation is a process dependent on the amount of plant material added each year and the type of base soil; both are affected by climate and the type of organisms present. Soils with humus can vary in nitrogen content but have 3 to 6 percent nitrogen typically; humus, as a reserve of nitrogen and phosphorus, is a vital component affecting soil fertility. Humus also absorbs water, acting as a moisture reserve, that plants can utilise; it also expands and shrinks between dry and wet states, providing pore spaces. Humus is less stable than other soil constituents, because it is affected by microbial decomposition, and over time its concentration decreases without the addition of new organic matter. However, some forms of humus are highly stable and may persist over centuries if not millennia: they are issued from the slow oxidation of charcoal, also called black carbon, like in Amazonian Terra preta or Black Earths, or from the sequestration of humic compounds within mineral horizons, like in podzols.

Climate and Organics

The production and accumulation or degradation of organic matter and humus is greatly dependent on climate conditions. Temperature and soil moisture are the major factors in the formation or degradation of organic matter, they along with topography, determine the formation of organic soils. Soils high in organic matter tend to form under wet or cold conditions where decomposer activity is impeded by low temperature or excess moisture.

Soil Solutions

Soils retain water that can dissolve a range of molecules and ions. These solutions exchange gases with the soil atmosphere, contain dissolved sugars, fulvic acids and other organic acids, plant nutrients such as nitrate, ammonium, potassium, phosphate, sulfate and calcium, and micronutrients such as zinc, iron and copper. These nutrients are exchanged with the mineral and humic component, that retains them in its ionic state, by adsorption. Some arid soils have sodium solutions that greatly impact plant growth. Soil pH can affect the type and amount of anions and cations that soil solutions contain and that be exchanged between the soil substrate and biological organisms.

In Nature

Biogeography is the study of special variations in biological communities. Soils determine which plants can grow in which environments. Soil scientists survey soils in the hope of understanding the parameters that determine what vegetation can and will grow in a particular location. Geologists also have a particular interest in the patterns of soil on the surface of the earth. Soil texture, colour and chemistry often reflect the underlying geologic parent material, and soil types often change at geologic unit boundaries. Buried paleosols mark previous land surfaces and record climatic conditions from previous eras. Geologists use this paleopedological record to understand the ecological relationships that existed in the past. According to the theory of biorhexistasy, prolonged conditions conducive to forming deep, weathered soils result in increasing ocean salinity and the formation of limestone.

Geologists use soil profile features to establish the duration of surface stability in the context of geologic faults or slope stability. An offset subsoil horizon indicates rupture during soil formation and the degree of subsequent subsoil formation is relied upon to establish time since rupture occurred.

Figure: A homeowner tests soil to apply only the nutrients needed.

Soil examined in shovel test pits is used by archaeologists for relative dating based on stratigraphy (as opposed to absolute dating). What is considered most typical is to use soil profile features to determine the maximum reasonable pit depth than needs to be examined for archaeological evidence in the interest of cultural resources management.

Figure: A homeowner sifts soil made from his compost bin in background. Composting is an excellent way to recycle organic wastes.

Soils altered or formed by humans (anthropic and anthropogenic soils) are also of interest to archaeologists, such as terra preta soils.

Uses

Soil is used in agriculture, where it serves as the anchor and primary nutrient base for plants; however, as demonstrated by hydroponics, it is not essential to plant growth if the soil-contained nutrients could be dissolved in a solution. The types of soil and available moisture determine the species of plants that can be cultivated.

Soil material is a critical component in the mining and construction industries. Soil serves as a foundation for most construction projects. The movement of massive volumes of soil can be involved in surface mining, road building and dam construction. Earth sheltering is the architectural practice of using soil for external thermal mass against building walls.

Soil resources are critical to the environment, as well as to food and fibre production. Soil provides minerals and water to plants. Soil absorbs rainwater and releases it later, thus preventing floods and

Introduction

drought. Soil cleans the water as it percolates through it. Soil is the habitat for many organisms: the major part of known and unknown biodiversity is in the soil, in the form of invertebrates (earthworms, woodlice, millipedes, centipedes, snails, slugs, mites, springtails, enchytraeids, nematodes, protists), bacteria, archaea, fungi and algae; and most organisms living above ground have part of them (plants) or spend part of their life cycle (insects) belowground. Above-ground and belowground biodiversities are tightly interconnected, making soil protection of paramount importance for any restoration or conservation plan.

The biological component of soil is an extremely important carbon sink since about 57% of the biotic content is carbon. Even on desert crusts, cyanobacteria lichens and mosses capture and sequester a significant amount of carbon by photosynthesis. Poor farming and grazing methods have degraded soils and released much of this sequestered carbon to the atmosphere. Restoring the world's soils could offset some of the huge increase in greenhouse gases causing global warming while improving crop yields and reducing water needs.

Waste management often has a soil component. Septic drain fields treat septic tank effluent using aerobic soil processes. Landfills use soil for daily cover. Land application of wastewater relies on soil biology to aerobically treat BOD.

Organic soils, especially peat, serve as a significant fuel resource; but wide areas of peat production, such as sphagnum bogs, are now protected because of patrimonial interest.

Both animals and humans in many cultures occasionally consume soil. It has been shown that some monkeys consume soil, together with their preferred food (tree foliage and fruits), in order to alleviate tannin toxicity.

Soils filter and purify water and affect its chemistry. Rain water and pooled water from ponds, lakes and rivers percolate through the soil horizons and the upper rock strata, thus becoming groundwater. Pests (viruses) and pollutants, such as persistent organic pollutants (chlorinated pesticides, polychlorinated biphenyls), oils (hydrocarbons), heavy metals (lead, zinc, cadmium), and excess nutrients (nitrates, sulfates, phosphates) are filtered out by the soil. Soil organisms metabolise them or immobilise them in their biomass and necromass, thereby incorporating them into stable humus. The physical integrity of soil is also a prerequisite for avoiding landslides in rugged landscapes.

Degradation

Land degradation is a human-induced or natural process which impairs the capacity of land to function. Soils are the critical component in land degradation when it involves acidification, contamination, desertification, erosion or salination.

While soil acidification of alkaline soils is beneficial, it degrades land when soil acidity lowers crop productivity and increases soil vulnerability to contamination and erosion. Soils are often initially acid because their parent materials were acid and initially low in the basic cations (calcium, magnesium, potassium and sodium). Acidification occurs when these elements are removed from the soil profile by normal rainfall, or the harvesting of forest or agricultural crops. Soil acidification is accelerated by the use of acid-forming nitrogenous fertilizers and by the effects of acid precipitation.

Soil contamination at low levels is often within soil capacity to treat and assimilate. Many waste treatment processes rely on this treatment capacity. Exceeding treatment capacity can damage soil biota and limit soil function.

Derelict soils occur where industrial contamination or other development activity damages the soil to such a degree that the land cannot be used safely or productively. Remediation of derelict soil uses principles of geology, physics, chemistry and biology to degrade, attenuate, isolate or remove soil contaminants to restore soil functions and values. Techniques include leaching, air sparging, chemical amendments, phytoremediation, bioremediation and natural attenuation.

Desertification is an environmental process of ecosystem degradation in arid and semi-arid regions, often caused by human activity. It is a common misconception that droughts cause desertification. Droughts are common in arid and semiarid lands. Well-managed lands can recover from drought when the rains return. Soil management tools include maintaining soil nutrient and organic matter levels, reduced tillage and increased cover. These practices help to control erosion and maintain productivity during periods when moisture is available. Continued land abuse during droughts, however, increases land degradation. Increased population and livestock pressure on marginal lands accelerates desertification.

Soil erosional loss is caused by wind, water, ice and movement in response to gravity. Although the processes may be simultaneous, erosion is distinguished from weathering. Erosion is an intrinsic natural process, but in many places it is increased by human land use. Poor land use practices including deforestation, overgrazing and improper construction activity.

Improved management can limit erosion by using techniques like limiting disturbance during construction, avoiding construction during erosion prone periods, intercepting runoff, terrace-building, use of erosion-suppressing cover materials, and planting trees or other soil binding plants.

A serious and long-running water erosion problem occurs in China, on the middle reaches of the Yellow River and the upper reaches of the Yangtze River. From the Yellow River, over 1.6-billion tons of sediment flow each year into the ocean. The sediment originates primarily from water erosion (gully erosion) in the Loess Plateau region of northwest China.

Soil piping is a particular form of soil erosion that occurs below the soil surface. It is associated with levee and dam failure, as well as sink hole formation. Turbulent flow removes soil starting from the mouth of the seep flow and subsoil erosion advances upgradient. The term sand boil is used to describe the appearance of the discharging end of an active soil pipe.

Soil salination is the accumulation of free salts to such an extent that it leads to degradation of soils and vegetation. Consequences include corrosion damage, reduced plant growth, erosion due to loss of plant cover and soil structure, and water quality problems due to sedimentation. Salination occurs due to a combination of natural and human caused processes.

Arid conditions favour salt accumulation. This is especially apparent when soil parent material is saline. Irrigation of arid lands is especially problematic. All irrigation water has some level of salinity. Irrigation, especially when it involves leakage from canals and overirrigation in the field, often raises the underlying water table. Rapid salination occurs when the land surface is within the capillary fringe of saline groundwater. Soil salinity control involves watertable control and flushing with higher levels of applied water in combination with tile drainage or another form of subsurface drainage.

Soil salinity models like SWAP, DrainMod-S, UnSatChem, SaltMod and SahysMod are used to assess the cause of soil salination and to optimise the reclamation of irrigated saline soils.

Soil-forming Materials

Rocks are the chief sources for the parent materials over which soils are developed. There are three main kinds of rocks:
(i) igneous rocks,
(ii) sedimentary rocks, and
(iii) metamorphic rocks.

Igneous rocks. They are formed by the cooling, hardening and crystallizing of various kinds of lavas and differ widely in their chemical composition. They chiefly contain feldspars, maphic minerals and quartz. Rocks containing a high proportion of quartz (60-75%) are classified as acidic, whereas those containing less than 50% quartz are classified as basic. The common igneous rocks found in India are the granites(acidic) and basalts or the Deccan Trap (basic)

Sedimentary rocks. They are derived from igneous rocks and are formed by the consolidation of fragmentary rock materials and the products of their decomposition deposited by water. The common sedimentary rocks are conglomerate, sandstone, shale and limestone. Alluvial, glacial and aeolian deposits form the unconsolidated sedimentary rocks.

Metamorphic rocks. They are formed from the igneous or sedimentary rocks by the action of intense heat and high pressure or both resulting in considerable change in the texture and mineral composition. The common metamorphic rocks are gneis from granite, quartzite from quartz or sandstone, marble from limestone and slate from shale.

The rocks vary greatly in chemical composition. Table1 gives the average composition of four different kinds of rocks

Table : Percentage Chemical Composition of Rocks

Oxides	Igneous rocks	Shales	Sandstones	Limestones
SiO_2	59.07	58.90	78.64	5.20
Al_2O_3	15.22	15.63	4.77	0.81
Fe_2O_3	3.10	4.07	1.08	0.54
MgO	3.45	2.47	1.17	7.92
FeO	3.71	2.48	0.32	..
CaO	5.10	3.15	5.51	42.74
Na_2O	3.71	1.32	0.45	0.05
K_2O	3.11	3.28	1.32	0.33

O_2	..	2.67	5.03	41.70
P_2O_5	0.30	0.17	.08	0.04
MnO	0.11	..	..	..
TiO_3	1.03	0.66	0.25	0.06
H_2O	1.30	3.72	1.33	0.56
Miscellaneous	0.79	1.48	0.07	0.05

Weathering refers to the physical and chemical disintegration and decomposition of rocks which are not under equilibrium under temperature, pressure and moisture conditions on the earth's surface. In the beginning, weathering precedes soil formation, more so in hard rocks. In other words, weathering creates the parent material over which soil formation takes place. Later, weathering, soil formation and development proceed simultaneously. The weathering may be physical or chemical.

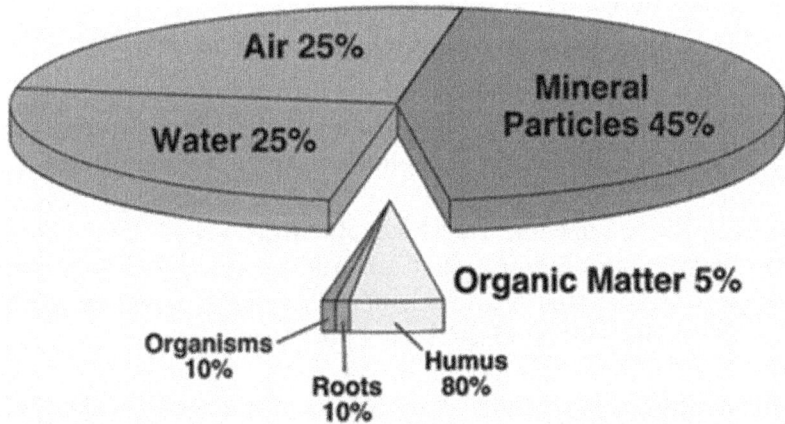

Figure: Most soils contain four basic components: mineral particles, water, air, and organic matter. Organic matter can be further subdivided into humus, roots, and living organisms. The values given above are for an average soil.

Soil itself is very complex. It would be very wrong to think of soils as just a collection of fine mineral particles. Soil also contains air, water, dead organic matter, and various types of living organisms. The formation of a soil is influenced by organisms, climate, topography, parent material, and time. The following items describe some important features of a soil that help to distinguish it from mineral sediments.

Organic Activity

A mass of mineral particles alone do not constitute a true soil. True soils are influenced, modified, and supplemented by living organisms. Plants and animals aid in the development of a soil through the addition of organic matter. Fungi and bacteria decompose this organic matter into a semi-soluble chemical substance known as humus. Larger soil organisms, like earthworms, beetles, and termites, vertically redistribute this humus within the mineral matter found beneath the surface of a soil. Humus is the biochemical substance that makes the upper layers of the soil become dark. It is coloured dark brown to black. Humus is difficult to see in isolation because it binds with larger mineral and organic particles. Humus provides soil with a number of very important benefits:

- It enhances a soil's ability to hold and store moisture.
- It reduces the eluviation of soluble nutrients from the soil profile.
- It is the primary source of carbon and nitrogen required by plants for their nutrition.
- It improves soil structure which is necessary for plant growth.

Organic activity is usually profuse in the near surface layers of a soil. For instance, one cubic centimetre of soil can be the home to more than 1,000,000 bacteria. A hectare of pasture land in a humid mid-latitude climate can contain more than a million earthworms and several million insects. Earthworms and insects are extremely important because of their ability mix and aerate soil. Higher porosity, because of mixing and aeration, increases the movement of air and water from the soil surface to deeper layers where roots reside. Increasing air and water availability to roots has a significant positive effect on plant productivity. Earthworms and insects also produce most of the humus found in a soil through the incomplete digestion of organic matter.

Translocation

When water moves downward into the soil, it causes both mechanical and chemical translocations of material. The complete chemical removal of substances from the soil profile is known as leaching. Leached substances often end up in the groundwater zone and then travel by groundwater flow into water bodies like rivers, lakes, and oceans. Eluviation refers to the movement of fine mineral particles (like clay) or dissolved substances out of an upper layer in a soil profile. The deposition of fine mineral particles or dissolved substances in a lower soil layer is called illuviation.

Soil Texture

The texture of a soil refers to the size distribution of the mineral particles found in a representative sample of soil. Particles are normally grouped into three main classes: sand, silt, and clay. The classification of soil particles according to size.

Table 1: Particle size ranges for sand, silt, and clay.

Type of Mineral Particle	*Size Range*
Sand	2.0 - 0.06 millimetres
Silt	0.06 - 0.002 millimetres
Clay	less than 0.002 millimetres

Clay is probably the most important type of mineral particle found in a soil. Despite their small size, clay particles have a very large surface area relative to their volume. This large surface is highly reactive and has the ability to attract and hold positively charged nutrient ions. These nutrients are available to plant roots for nutrition. Clay particles are also somewhat flexible and plastic because of their lattice-like design. This feature allows clay particles to absorb water and other substances into their structure.

Soil pH

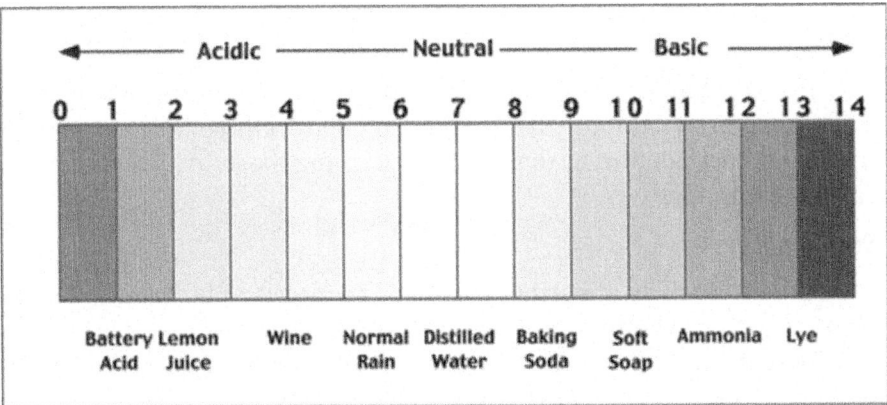

Figure: The pH scale. A value of 7.0 is considered neutral. Values higher than 7.0 are increasingly alkaline or basic. Values lower than 7.0 are increasingly acidic. The illustration above also describes the pH of some common substances.

Soils support a number of inorganic and organic chemical reactions. Many of these reactions are dependent on some particular soil chemical properties. One of the most important chemical properties influencing reactions in a soil is pH. Soil pH is primarily controlled by the concentration of free hydrogen ions in the soil matrix. Soils with a relatively large concentration of hydrogen ions tend to be acidic. Alkaline soils have a relatively low concentration of hydrogen ions. Hydrogen ions are made available to the soil matrix by the dissociation of water, by the activity of plant roots, and by many chemical weathering reactions.

Soil fertility is directly influenced by pH through the solubility of many nutrients. At a pH lower than 5.5, many nutrients become very soluble and are readily leached from the soil profile. At high pH, nutrients become insoluble and plants cannot readily extract them. Maximum soil fertility occurs in the range 6.0 to 7.2.

Soil Colour

Soils tend to have distinct variations in colour both horizontally and vertically. The colouring of soils occurs because of a variety of factors.

Soils of the humid tropics are generally red or yellow because of the oxidation of iron or aluminium, respectively. In the temperate grasslands, large additions of humus cause soils to be black. The heavy leaching of iron causes coniferous forest soils to be gray. High water tables in soils cause the reduction of iron, and these soils tend to have greenish and gray-blue hues. Organic matter colours the soil black.

The combination of iron oxides and organic content gives many soil types a brown colour. Other colouring materials sometimes present include white calcium carbonate, black manganese oxides, and black carbon compounds.

Soil Profiles

Most soils have a distinct profile or sequence of horizontal layers. Generally, these horizons result from the processes of chemical weathering, eluviation, illuviation, and organic decomposition. Up to five layers can be present in a typical soil: O, A, B, C, and R horizons.

The O horizon is the topmost layer of most soils. It is composed mainly of plant litter at various levels of decomposition and humus.

A horizon is found below the O layer. This layer is composed primarily of mineral particles and has two characteristics: it is the layer in which humus and other organic materials are mixed with mineral particles, and it is a zone of translocation from which eluviation has removed finer particles and soluble substances, both of which may be deposited at a lower layer. Thus the A horizon is dark in colour and usually light in texture and porous. The A horizon is commonly differentiated into a darker upper horizon or organic accumulation, and a lower horizon showing loss of material by eluviation.

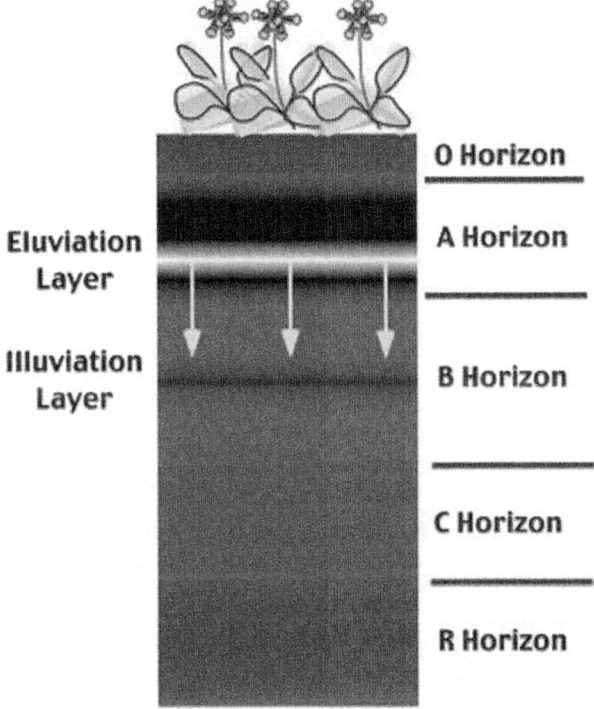

Figure: Typical layers found in a soil profile. (Source: PhysicalGeography.net)

The B horizon is a mineral soil layer which is strongly influenced by illuviation. Consequently, this layer receives material eluviated from the A horizon. The B horizon also has a higher bulk density than the A horizon due to its enrichment of clay particles. The B horizon may be coloured by oxides of iron and aluminium or by calcium carbonate illuviated from the A horizon.

The C horizon is composed of weathered parent material. The texture of this material can be quite variable with particles ranging in size from clay to boulders. The C horizon has also not been significantly influenced by the pedogenic processes, translocation, and/or organic modification.

The final layer in a typical soil profile is called the R horizon. This soil layer simply consists of unweathered bedrock.

Soil Pedogenesis

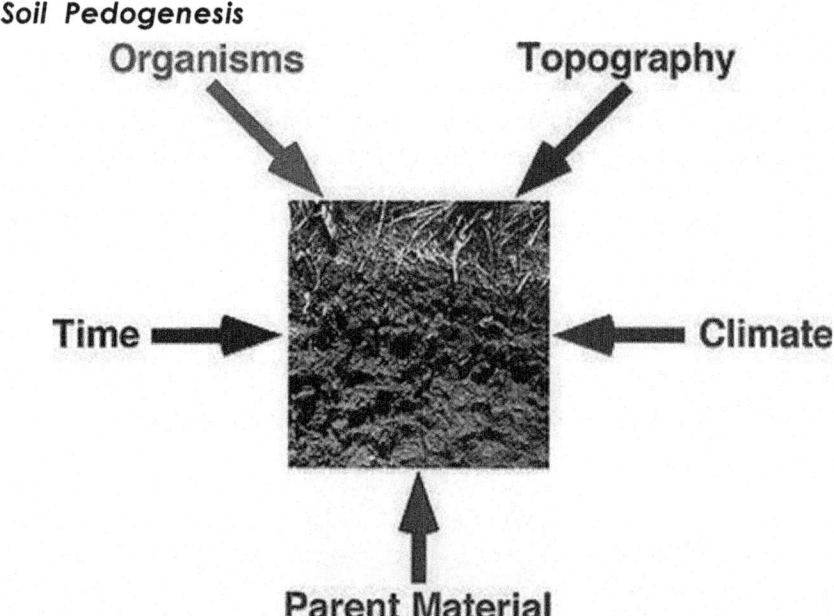

Figure: The development of a soil is influenced by five interrelated factors: organisms, topography, time, parent material, and climate.

Pedogenesis can be defined as the process of soil development. Late in the 19th century, scientists Hilgard in the United States and the Russian Dukuchaev both suggested independently that pedogenesis was principally controlled by climate and vegetation. This idea was based on the observation that comparable soils developed in spatially separate areas when their climate and vegetation were similar. In the 1940s, Hans Jenny extended these ideas based on the observations of many subsequent studies examining the processes involved in the formation of soils. Jenny believed that the kinds of soils that develop in a particular area are largely determined by five interrelated factors: climate; living organisms; parent material; topography; and time.

Climate plays a very important role in the genesis of a soil. On the global scale, there is an obvious correlation between major soil types and the Köppen climatic classification systems major climatic types. At regional and local scales, climate becomes less important in soil formation. Instead, pedogenesis is more influenced by factors like parent material, topography, vegetation, and time. The two most important climatic variables influencing soil formation are temperature and moisture. Temperature has a direct influence on the weathering of bedrock to produce mineral particles. Rates of bedrock weathering generally increase with higher temperatures. Temperature also influences the activity of soil microorganisms, the frequency and magnitude of soil chemical reactions, and the rate of plant growth. Moisture levels in most soils are primarily controlled by the addition of water via precipitation minus the losses due to evapotranspiration. If additions of water from precipitation surpass losses from evapotranspiration, moisture levels in a soil tend to be high. If the water loss due to evapotranspiration exceeds inputs from precipitation, moisture levels in a soil tend to be low. High moisture availability in a soil promotes the weathering of bedrock and sediments, chemical reactions, and plant growth. The availability of moisture also has an influence on soil pH and the decomposition of organic matter.

Living organisms have a role in a number of processes involved in pedogenesis including organic matter accumulation, profile mixing, and biogeochemical nutrient cycling. Under equilibrium conditions, vegetation and soil are closely linked with each other through nutrient cycling. The cycling of nitrogen and carbon in soils is almost completely controlled by the presence of animals and plants. Through litterfall and the process of decomposition, organisms add humus and nutrients to the soil which influences soil structure and fertility. Surface vegetation also protects the upper layers of a soil from erosion by way of binding the soils surface and reducing the speed of moving wind and water across the ground surface.

Parent material refers to the rock and mineral materials from which the soils develop. These materials can be derived from residual sediment due to the weathering of bedrock or from sediment transported into an area by way of the erosive forces of wind, water, or ice. Pedogenesis is often faster on transported sediments because the weathering of parent material usually takes a long period of time. The influence of parent material on pedogenesis is usually related to soil texture, soil chemistry, and nutrient cycling.

Topography generally modifies the development of soil on a local or regional scale. Pedogenesis is primarily influenced by topography's effect on microclimate and drainage. Soils developing on moderate to gentle slopes are often better drained than soils found at the bottom of valleys. Good drainage enhances an number of pedogenic processes of illuviation and eluviation that are responsible for the development of soil horizons. Under conditions of poor drainage, soils tend to be immature. Steep topographic gradients inhibit the development of soils because of erosion. Erosion can retard the development through the continued removal of surface sediments. Soil microclimate is also influenced by topography. In the Northern Hemisphere, south-facing slopes tend to be warmer and drier than north-facing slopes. This difference results in the soils of the two areas being different in terms of depth, texture, biological activity, and soil profile development.

Time influences the temporal consequences of all of the factors described above. Many soil processes become steady state overtime when a soil reaches maturity. Pedogenic processes in young soils are usually under active modification through negative and positive feedback mechanisms in attempt to achieve equilibrium.

Principal Pedogenic Processes

A large number of processes are responsible for the formation of soils. This fact is evident by the large number of different types of soils that have been classified by soil scientists. However, at the macro-scale we can suggest that there are five main principal pedogenic processes acting on soils. These processes are laterization, podzolization, calcification, salinization, and gleization.

Laterization is a pedogenic process common to soils found in tropical and subtropical environments. High temperatures and heavy precipitation result in the rapid weathering of rocks and minerals. Movements of large amounts of water through the soil cause eluviation and leaching to occur. Almost all of the byproducts of weathering, very simple small compounds or nutrient ions, are translocated out of the soil profile by leaching if not taken up by plants for nutrition. The two exceptions to this process are iron and aluminium compounds. Iron oxides give tropical soils their unique reddish colouring. Heavy leaching also causes these soils to have an acidic pH because of the net loss of base cations.

Podzolization is associated with humid cold mid-latitude climates and coniferous vegetation. Decomposition of coniferous litter and heavy summer precipitation create a soil solution that is strongly acidic. This acidic soil solution enhances the processes of eluviation and leaching causing the removal of soluble base cations and aluminium and iron compounds from the A horizon. This process creates a sub-layer in the A horizon that is white to gray in colour and composed of silica sand.

Calcification occurs when evapotranspiration exceeds precipitation causing the upward movement of dissolved alkaline salts from the groundwater. At the same time, the movement of rain water causes a downward movement of the salts. The net result is the deposition of the translocated cations in the B horizon. In some cases, these deposits can form a hard layer called caliche. The most common substance involved in this process is calcium carbonate. Calcification is common in the prairie grasslands.

Salinization is a process that functions in the similar way to calcification. It differs from calcification in that the salt deposits occur at or very near the soil surface. Salinization also takes place in much drier climates.

Gleization is a pedogenic process associated with poor drainage. This process involves the accumulations of organic matter in the upper layers of the soil. In lower horizons, mineral layers are stained blue-gray because of the chemical reduction of iron.

Soil Classification

Soil Classification Systems have been developed to provide scientists and resource managers with generalised information about the nature of a soil found in a particular location. In general, environments that share comparable soil-forming factors produce similar types of soils. This phenomenon makes classification possible. Numerous classification systems are in use worldwide. We will examine the systems commonly used in the United States and Canada.

United States Soil Classification System

The first formal system of soil classification was introduced in the United States by Curtis F. Marbut in the 1930s. This system, however, had some serious limitations, and by the early 1950s the United States Soil Conservation Service (now the Natural Resources Conservation Service) began the development of a new method of soil classification. The process of development of the new system took nearly a decade to complete. By 1960, the review process was completed and the Seventh Approximation Soil Classification System was introduced. Since 1960, this soil classification system has undergone numerous minor modifications and is now under the control of Natural Resources Conservation Service (NRCS), which is a branch of the Department of Agriculture. The current version of the system has six levels of classification in its hierarchical structure. The major divisions in this classification system, from general to specific, are: orders, suborders, great groups, subgroups, families, and series. At its lowest level of organisation, the U.S. system of soil classification recognises approximately 15,000 different soil series.

The most general category of the NRCS Soil Classification System recognises eleven distinct soil orders: oxisols, aridsols, mollisols, alfisols, ultisols, spodsols, entisols, inceptisols, vertisols, histosols, and andisols.

Oxisols develop in tropical and subtropical latitudes that experience an environment with high precipitation and temperature. The profiles of oxisols contain mixtures of quartz, kaolin clay, iron and aluminium oxides, and organic matter.

For the most part they have a nearly featureless soil profile without clearly marked horizons. The abundance of iron and aluminium oxides found in these soils results from strong chemical weathering and heavy leaching. Many oxisols contain laterite layers because of a seasonally fluctuating water table.

Aridsols are soils that develop in very dry environments. The main characteristic of this soil is poor and shallow soil horizon development. Aridsols also tend to be light-coloured because of limited humus additions from vegetation. The hot climate under which these soils develop tends to restrict vegetation growth. Because of limited rain and high temperatures soil water tends to migrate in these soils in an upward direction. This condition causes the deposition of salts carried by the water at or near the ground surface because of evaporation. This soil process is called salinization.

Mollisols are soils common to grassland environments. In the United States most of the natural grasslands have been converted into agricultural fields for crop growth. Mollisols have a dark-coloured surface horizon, tend to be base rich, and are quite fertile. The dark colour of the A horizon is the result of humus enrichment from the decomposition of litterfall. Mollisols found in more arid environments often exhibit calcification.

Alfisols form under forest vegetation where the parent material has undergone significant weathering. These soils are quite widespread in their distribution and are found from southern Florida to northern Minnesota. The most distinguishing characteristics of this soil type are the illuviation of clay in the B horizon, moderate to high concentrations of base cations, and light-coloured surface horizons.

Ultisols are soils common to the southeastern United States. This region receives high amounts of precipitation because of summer thunderstorms and the winter dominance of the mid-latitude cyclone. Warm temperatures and the abundant availability of moisture enhances the weathering process and increases the rate of leaching in these soils. Enhanced weathering causes mineral alteration and the dominance of iron and aluminium oxides. The presence of the iron oxides causes the A horizon of these soils to be stained red. Leaching causes these soils to have low quantities of base cations.

Spodsols are soils that develop under coniferous vegetation and as a result are modified by podzolization. Parent materials of these soils tend to be rich in sand.

The litter of the coniferous vegetation is low in base cations and contributes to acid accumulations in the soil. In these soils, mixtures of organic matter and aluminium, with or without iron, accumulate in the B horizon. The A horizon of these soils normally has an eluvial layer that has the colour of more or less quartz sand. Most spodosols have little silicate clay and only small quantities of humus in their A horizon. Entisols are immature soils that lack the vertical development of horizons. These soils are often associated with recently deposited sediments from wind, water, or ice erosion. Given more time, these soils will develop into another soil type.

Inceptisols are young soils that are more developed than entisols. These soils are found in arctic tundra environments, glacial deposits, and relatively recent deposits of stream alluvium. Common characteristics of recognition include immature development of eluviation in the A horizon and illuviation in the B horizon, and evidence of the beginning of weathering processes on parent material sediments.

Vertisols are heavy clay soils that show significant expansion and contraction due to the presence or absence of moisture. Vertisols are common in areas that have shale parent material and heavy precipitation. The location of these soils in the United States is primarily found in Texas where they are used to grow cotton.

Histosols are organic soils that form in areas of poor drainage. Their profile consists of thick accumulations of organic matter at various stages of decomposition.

Andisols develop from volcanic parent materials. Volcanic deposits have a unique process of weathering that causes the accumulation of allophane and oxides of iron and aluminium in developing soils.

Canadian System of Soil Classification

Canada's first independent taxonomic system of soil classification was first introduced in 1955. Prior to 1955, systems of classification used in Canada were strongly based on methods being applied in the United States. However, the U.S. system was based on environmental conditions common to the United States. Canadian soil scientists required a new method of soil classification that focused on pedogenic processes in cool climatic environments.

Like the U.S. system, the Canadian System of Soil Classification differentiates soil types on the basis of measured properties of the profile and uses a hierarchical scheme to classify soils from general to specific. The most recent version of the classification system has five categories in its hierarchical structure. From general to specific, the major categories in this system are: orders, great groups, subgroups, families, and series. At its most general level, the Canadian System recognises nine different soil orders:

- Brunisol - is a normally immature soil commonly found under forested ecosystems. The most identifying trait of these soils is the presence of a B horizon that is brownish in colour. The soils under the dry pine forests of south-central British Columbia are typically brunisols.
- Chernozem - is a soil common to grassland ecosystems. This soil is dark in colour (brown to black) and has an A horizon that is rich in organic matter. Chernozems are common in the Canadian prairies. The images below are from the eastern prairies where higher seasonal rainfalls produce black chernozemic soils.
- Cryosol - is a high-latitudes soil common in the tundra. This soil has a layer of permafrost within one metre of the soil surface. The soil profile has a permanently frozen ice wedge beneath its surface.
- Gleysol - is a soil found in an ecosystem that is frequently flooded or permanently waterlogged. Its soil horizons show the chemical signs of oxidation and reduction.
- Luvisol - is another type of soil that develops under forested conditions. This soil, however, has a calcareous parent material which results in a high pH and strong eluviation of clay from the A horizon.

- Organic - this soil is mainly composed of organic matter in various stages of decomposition. Organic soils are common in fens and bogs. The profiles of these soils have an obvious absence of mineral soil particles.
- Podzol - is a soil commonly found under coniferous forests. Its main identifying traits are a poorly decomposed organic layer, an eluviated A horizon, and a B horizon with illuviated organic matter, aluminium, and iron. The forested regions of southern Ontario and the temperate rainforests of British Columbia normally have podzolic soils.
- Regosol - is any young underdeveloped soil. Immature soils are common in geomorphically dynamic environments. Many mountain river valleys in British Columbia have floodplains with surface deposits that are less than 3,000 years old. The soils in these environments tend to be regosols.
- Solonetzic - is a grassland soil where high levels of evapotranspiration cause the deposition of salts at or near the soil surface. Solonetzic soils are common in the dry regions of the prairies where evapotranspiration greatly exceeds precipitation input. The movement of water to the earth's surface because of capillary action, transpiration, and evaporation causes the deposition of salts when the water evaporates into the atmosphere.

Soils of India

Types of Soil Found in India

Indian soils are generally divided into four broad types. These soil types are: 1) alluvial soils; 2) regur soils; 3) red soils and 4) laterite soils.

Alluvial Soils: This is the most important and widespread category. It covers 40% of the land area. In fact the entire Northern Plains are made up of these soils. They have been brought down and deposited by three great Himalayan rivers- Sutlej, Ganga and Brahmaputra- and their tributaries. Through a narrow corridor in Rajasthan they extend into the plains of Gujarat. They are common in eastern coastal plains and in the deltas of Mahanadi, Godavari, Krishna and Kaveri.

Regur Soils: These soils are black in colour and are also known as black soils. Since, they are ideal for growing cotton, they are also called cotton soils, in addition to their normal nomenclature of regur soils. These soils are most typical of the Deccan trap (Basalt) region spread over north-west Deccan plateau and are made up of lava flows. They cover the plateaus of Mahrashtra, Saurashtra, Malwa and southern Madhya Pradesh and extends eastwards in the south along the Godavari and Krishna Valleys.

Red Soils: These soils are developed on old crystalline rocks under moderate to heavy rainfall conditions. They are deficient in phosphoric acid, organic matter and nitrogenous material. Red soils cover the eastern part of the peninsular region comprising Chhotanagpur plateau, Orissa, eastern Madhya Pradesh, Telangana, the Nilgiris and Tamil Nadu plateau. Tey extended northwards in the west along the Konkan coast of Maharashtra.

Laterite Soils: The laterite soils is the result of intense leaching owing to heavy tropical rains. They are found along the edge of plateau in the east covering small parts of Tamil Nadu, and Orissa and a small part of Chhotanagpur in the north and Meghalaya in the north-east.

Important facts about the soil:
- Types of soil: -Clay, silt and sand are the three types of soil. Most soils are a blend of all three types. The texture and appearance of soil depends on the content of each of these. Sand is mainly granular and is composed of rock particles and minerals. Clay has fine-grained minerals and high water content. Silt is a granular material derived from rock. It may occur as a deposition in water. Silt is also known as stone-dust.
- Composition of soil: Soil holds 0.01% of the Earth's water. Soil is a composition of 49% Oxygen, 33% Silicone, 7% Aluminium, 4% Iron, and 2% Carbon. Air and water make up 50% of the soil. Minerals and organic matter make up the rest.
- Formation of soil: Soil formation is a lengthy process. Soil forms by the process of physical or chemical weathering of rocks. Microorganisms in the soil help in breakdown of organic matter in the soil. Decaying of plants and animals helps in the formation of soil. Earthworms recycle nutrients thus making the soil richer.
- Layers of soil: -The topmost layer of soil is called topsoil. It contains high amounts of humus and microorganisms. Biological activity occurs most in this layer. It is from this layer that

plants derive their nutrients. Not much humus is present in the layer below this layer. The process of leaching brings down the minerals from the upper layers to the layers below. The bottom-most layer consists of withered rock.

Interesting Facts about the soil:
- Soil influences many areas of our lives. It is an integral part of our ecosystem. The composition of the soil in an area has a direct effect on the plant and animal life there.
- It takes more than 500 years to form 2 centimetres of topsoil.
- Ten tons of topsoil spread evenly over one hectare of land comes out to be as thick as one Euro coin.
- A fully functional soil holds 3750 tons of water per hectare, thus reducing the risk of floods. It holds pollutants to a certain extent. Soil stores around 10% of the emissions of carbon dioxide.
- Just one gram of soil contains 5000 to 7000 different species of bacteria. A spoonful of soil can hold a substantial amount of living beings.
- Scientists have found 10,000 types of soil in Europe and about 70,000 types of soil in the United States.
- 75% of the earth's crust is composed of silica and oxygen.
- Soil is a non-renewable natural resource. This should make us think of how much we value this resource. Damage to the soil can disturb nature's balance and prove a threat to life.

It is in this soil that crops grow and we can obtain our food. Many of the antibiotics that stand as remedies for infections, were obtained from microorganisms in the soil. As a matter of fact, agriculture remains to be the only essential industry. Soil in its various forms plays a major role in our lives. In the words of the Greek philosopher, poet Xenophanes, "For all things come from earth, and all things end by becoming earth."

Different Types of Soil

Soil is the thin layer on the surface of the Earth on which the living beings of the earth survive since it is the layer of materials in which plants have their roots. Soil is made up of many things like weathered rock particles and decayed plant and animal matter. It takes a long time for soil formation and more than thousand years for the formation of a thin layer of soil. Since soil is made up of such diverse materials like broken down rock particles and organic material, it can be classified into various types, though based on the size of the particles it contains.

Soil Types

Therefore depending on the size of the particles in the soil, it can be classified into these following types:
- Sandy soil
- Silty soil
- Clay soil
- Loamy Soil
- Peaty Soil
- Chalky Soil.

Sandy Soil

This type has the biggest particles and the size of the particles does determine the degree of aeration and drainage that the soil allows. It is granular and consists of rock and mineral particles that are very small. Therefore the texture is gritty and sandy soil is formed by the disintegration and weathering of rocks such as limestone, granite, quartz and shale.

Sandy soil is easier to cultivate if it is rich in organic material but then it allows drainage more than is needed, thus resulting in over-drainage and dehydration of the plants in summer. It warms very fast in the spring season. So if you want to grow your plant in sandy soil it is imperative that you water it regularly in the summers and give a break in the winters and rainy season, sandy soil retains moisture and nutrients. In a way sandy soil is good for plants since it lets the water go off so that it does not remain near the roots and lead them to decay.

Silty Soil

Silty soil is considered to be one of the most fertile of soils. It can occur in nature as soil or as suspended sediment in water column of a water body on the surface of the earth. It is composed of minerals like Quartz and fine organic particles. It is granular like sandy soil but it has more nutrients than sandy soil and it also offers better drainage. In case silty soil is dry it has a smoother texture and looks like dark sand. This type of soil can hold more moisture and at times becomes compact. It offers better drainage and is much easier to work with when it has moisture.

Clay Soil

Clay is a kind of material that occurs naturally and consists of very fine grained material with very less air spaces, that is the reason it is difficult to work with since the drainage in this soil is low, most of the time there is a chance of water logging and harm to the roots of the plant. Clay soil becomes very heavy when wet and if cultivation has to be done, organic fertilizers have to be added. Clay soil is formed after years of rock disintegration and weathering. It is also formed as sedimentary deposits after the rock is weathered, eroded and transported.

Loamy Soil

This soil consists of sand, silt and clay to some extent. It is considered to be the perfect soil. The texture is gritty and retains water very easily, yet the drainage is well. There are various kinds of loamy soil ranging from fertile to very muddy and thick sod. Yet out of all the different kinds of soil loamy soil is the ideal for cultivation.

Peaty Soil

This kind of soil is basically formed by the accumulation of dead and decayed organic matter, it naturally contains much more organic matter than most of the soils. It is generally found in marshy areas. Now the decomposition of the organic matter in Peaty soil is blocked by the acidity of the soil. This kind of soil is formed in wet climate. Though the soil is rich in organic matter, nutrients present are fewer in this soil type than any other type. Peaty soil is prone to water logging but if the soil is fertilized well and the drainage of the soil is looked after, it can be the ideal for growing plants.

Chalky Soil

Unlike Peaty soil, Chalky soil is very alkaline in nature and consists of a large number of stones. The fertility of this kind of soil depends on the depth of the soil that is on the bed of chalk.

This kind of soil is prone to dryness and in summers it is a poor choice for plantation, as the plants would need much more watering and fertilizing than on any other type of soil. Chalky Soil, apart from being dry also blocks the nutritional elements for the plants like Iron and Magnesium.

Besides this kind of classification soil can also be classified as Acidic and Alkaline soil depending on the amount of humus, organic matter and the underlying bedrock. Every soil has its own advantages and disadvantages and there are various plants that have different requirements. All plants do not need the same kind of soil.

Humus Soil

Decomposing organic matter in the humus soil contributes to its fertility, owing to which it is considered to be the best bet by the stalwarts in the field of agriculture. Continue reading for more information on the various aspects of humus soil.

Humus soil is considered to be one of the most fertile types of soil in the world, and that explains its wide use for planting across the globe. In fact, humus soil is so fertile that it's widely used to treat other soil types as well. Going by the simplest humus soil definition, it is the soil with substantial amount of decomposed matter in it. However, this is just the basic aspect, as there is a lot more to know about this type of soil.

What is Humus Soil?

Humus soil is the soil which is produced over a period of time as a result of the decomposing organic matter. This organic matter may include anything, ranging from fallen leaves to animal waste. As this organic matter decomposes, it tends to form tiny negatively charged humus particles in the soil. The presence of these negatively charged particles in the soil in turns out to be beneficial as it absorbs the positively charged nutrients, such as calcium and magnesium, improves the fertility of the soil. As far as the appearance is concerned, humus soil ranges from dark brown to black in colour, with specks of white in it.

The process by which this soil is formed is referred to as humification. This process can happen naturally, i.e. on its own, or artificially, i.e. by the means of composting. An example of natural humification would be the process wherein the leaves that are shed by the plants start decaying, and the decayed matter is added to the soil, thus enhancing its fertility. A similar process occurs when animal waste is added to the soil. Read more on soil science.

Humus Soil Fertility

The organic matter, such as decaying plants, soil organisms or animal waste, adds to the fertility of this soil type, and thus makes it one of the most fertile soils, ideal for plant growth. This soil has the capacity to hold the mineral particles together in clusters referred to as the aggregates. These aggregates, in turn, improve the structure of the soil, and contribute to its fertility. Even though you can opt to buy humus soil for your garden, you can go for this option only when you don't have time to spare, as making humus soil is a relatively easy task.

How to Make Humus Soil

If you have an hour a day to spare, you can prepare humus soil on your own, through the method of composting. You can prepare humus soil using yard debris, such as fallen leaves or mowed grass, as well as your household waste. If required, you can also add animal waste to it. You just have to add this organic matter to the soil, and allow it to decompose in the compost pit.

You will have to turn the matter at least once a week to ensure that there is enough oxygen to facilitate the process of humification. Similarly, you will also have to monitor the moisture content of the soil. After some time, you will see tiny white matter in this soil, which would be a sign of your humus soil being ready to use. Using this method to prepare humus soil on your own will also ensure that you have it in abundance all round the year.

You should also go through:
- Facts About Soil
- Different Types of Soil.

This was brief information about humus soil, and how to prepare humus rich soil on your own. The rate at which the organic matter is converted to humus plays a crucial role in the life of this soil. More importantly, the presence of humus in soil also alters its ability to withstand drought conditions. At the end of the day, the overall verdict is that the humus soil is the best bet, if you are planning to start gardening, irrespective of whether you buy it or make it on your own.

Soil Pollution Causes and Effects

This article serves all our concerned and responsible readers with soil pollution causes and effects. This is because along with air and water pollution, soil pollution is an equally serious issue that the modern-day world is confronted with.

Soil Pollution

One of the most grave problems existing on the earth. Well, (on the earth?), the earth itself is getting contaminated and polluted! Collectively, aren't we all responsible for this? The conquest of utilising land and soil resources and conducting experiments on it, for our benefits, is quite understandable, but it certainly is not, at the cost of its health and wellness!

Mankind has been trying out several different things and has made several arrangements in the soil, to make life happy and comfortable. However, how often have we thought of contamination of soil? It's never too late in life, so I think this is the right time, to know about soil pollution causes and effects.

What is soil pollution? Soil pollution is defined or can be described as the contamination of soil of a particular region. Soil pollution mainly is a result of penetration of harmful pesticides and insecticides, which on one hand serve whatever their main purpose is, but on the other hand, bring about deterioration in the soil quality, thus making it contaminated and unfit for use.

Insecticides and pesticides are not to be blamed alone for soil pollution, but there are many other leading causes of soil pollution too. Let us have a look at some of them in the following text.

What Causes Soil Pollution?

Soil pollution is a result of many activities and experiments done by mankind and some of the leading soil pollution causes are discussed below.

- Industrial wastes, such as harmful gases and chemicals, agricultural pesticides, fertilizers and insecticides are the most important causes of soil pollution.
- Ignorance towards soil management and related systems.
- Unfavourable and harmful irrigation practices.
- Improper septic system and management and maintenance of the same.
- Leakages from sanitary sewage.
- Acid rains, when fumes released from industries get mixed with rains.
- Fuel leakages from automobiles, that get washed away due to rain and seep into the nearby soil.
- Unhealthy waste management techniques, which are characterised by release of sewage into the large dumping grounds and nearby streams or rivers.

The intensity of all these causes on a local or regional level might appear very small and you may argue that soil is not harmed by above activities if done on a small scale! However, thinking globally, it is not your region or my place, that will be the only sufferer of soil pollution. In fact, it is the entire planet and mankind that will encounter serious problems, as these practices are evident almost everywhere in the world. Want to know what are those problems, which can turn more serious in the near future?

What are the Effects of Soil Pollution?

The effects of pollution on soil are quite alarming and can cause huge disturbances in the ecological balance and health of living creatures on earth. Some of the most serious soil pollution effects are mentioned below.

- Decrease in soil fertility and therefore decrease in the soil yield. Definitely, how can one expect a contaminated soil to produce healthy crops?
- Loss of soil and natural nutrients present in it. Plants also would not thrive in such a soil, which would further result in soil erosion.

- Disturbance in the balance of flora and fauna residing in the soil.
- Increase in salinity of the soil, which therefore makes it unfit for vegetation, thus making it useless and barren.
- Generally crops cannot grow and flourish in a polluted soil. Yet if some crops manage to grow, they would be poisonous enough to cause serious health problems in people consuming them.
- Creation of toxic dust leading is another potential effect of soil pollution.
- Foul smell due to industrial chemicals and gases might result in headaches, fatigue, nausea, etc. in many people.
- Soil pollutants would bring in alteration in the soil structure, which would lead to death of many essential organisms in it. This would also affect the larger predators and compel them to move to other places, once they lose their food supply.

I hope the above discussion was enough to make you understand the severity of the soil pollution causes and effects. Soil pollution can be cured by transporting the contaminated soil layer to some remote place, thus making it once again fit for use.

Harmful chemicals from the soil can also be removed by aerating it. These are just 'tentative solutions'.

However, let us remember the proverb, 'prevention is better than cure' and follow soil management system, maintain sewage systems and avoid the overuse of fertilizers and pesticides in the soil. So let us begin the movement of soil pollution prevention from our own lands itself!!!

Soil Moisture Levels

Measuring soil moisture level is important for many hydrological, biological and biogeochemical processes. To know more about soil moisture levels measuring and monitoring equipment, read on...

The water particles held in the gaps between the soil molecules are called soil moisture, in soil science. Soil moisture measuring is important, because it helps in irrigation scheduling, estimating crop yields, and reservoir management.

Soil moisture measurement gives us an idea about soil erosion, slope failure and the quality of the water. It also gives early warning of droughts and floods. Soil moisture helps to control water evaporation and heat energy released through plant transpiration between the soil surface and the atmosphere.

It is a potent variable in the downfall of water and the development of weather patterns. Agencies and companies, that are concerned with water and weather, use soil moisture equipment for measuring and monitoring.

How to Measure Soil Moisture

There are various types of soils and each of them has a different capacity to hold water. This capacity depends upon the parameters, such as structure and texture of the soil, vegetation in the area, etc. Soil moisture level can be estimated by the feel and appearance of the soil, but soil moisture equipment gives accurate results. Nowadays, various measuring and monitoring methods and equipment are widely used for this purpose.

Soil Moisture Metres

Various soil moisture metres, such as tensiometres, equitensiometres and moisture metres are available in the market. They are simple to use and give accurate results.

Tensiometre

The tensiometre is used for the estimation of soil water suction. It is a sealed tube, which is filled with water. It has a special ceramic, porous tip and vacuum gauge on the upper end. Its length ranges from 6 inches to 72 inches. It is used to determine the irrigation needs, when the soil moisture is above 50% of the field capacity.

Neutron Moisture Metre

In a neutron moisture metre, state-of-the-art electronics and radioactive sources are used. As the radioactive source can be harmful for health, using the neutron moisture metre requires a license. It gives accurate results, but it is an expensive equipment.

Soil Moisture Sensors

Sensors are usually buried in the soil. They are either hard-wired to a fixed metre or they may have long attached electrodes, which are placed on the soil surface and hooked to a portable metre. Recording readings and plotting charts increases the precision of these sensors. For soil moisture sensing, electrical resistance blocks are most commonly used.

Soil Moisture Monitoring

Soil moisture monitoring techniques are easy to learn and they are also cost-efficient. These monitoring methods help to optimise crop yields, conserve water and energy, prevent soil erosion and water pollution. Soil moisture control, monitoring, prediction and irrigation scheduling are extremely important. Various probes and loggers are used for soil moisture monitoring.

Soil Moisture Probes

Probes are automatic, electronically-operated soil monitoring devices. They are very useful in environmental monitoring, irrigation, rain monitoring, sprinkler systems, weather monitoring, moisture monitoring of bulk foods, water conservation applications and fluid level measurements.

These soil moisture equipment are very helpful in biofuel studies, archeology, erosion studies, drought and flood forecasting models, landslide studies. They are also used in phytoremediation, which is a process, in which soil or water is decontaminated by using plants and trees to, either absorb or break down pollutants. Thus, soil moisture levels help in ascertaining various geological, biological and hydrological studies.

Soil Electrical Conductivity Variability

Soil electrical conductivity (EC) is a property of soil that is determined by standardized measures of soil conductance (resistance[1]) by the distance and cross sectional area through which a current travels. Traditionally, soil paste EC has been used to assess soil salinity (Rhoades et al., 1989), but now commercial devices are available to rapidly and economically measure and map bulk soil EC across agricultural fields.

The Veris® 3100 (Veris Technologies, Salina, Kansas) measures EC with a system of coulters that are in direct contact with the soil. The EM38 (Geonics, Limited, Mississauga, Ontario, Canada) induces a current into the soil with one coil and determines conductivity by measuring the resulting secondary current with another coil. Both sensors have been demonstrated to give similar results (Suddeth et al., 1999)

The movement of electrons through bulk soil is complex. Electrons may travel through soil water in macropores, along the surfaces of soil minerals (i.e. exchangeable ions), and through alternating layers of particles and solution (Rhoades et al., 1989).

Therefore, multiple factors contribute to soil EC variability, including factors that affect the amount and connectivity of soil water (e.g. bulk density, structure, water potential, precipitation, timing of measurement), soil aggregation (e.g. cementing agents such as clay and organic matter, soil structure), electrolytes in soil water (e.g. salinity, exchangeable ions, soil water content, soil temperature), and the conductivity of the mineral phase (e.g. types and quantity of minerals, degree of isomorphic substitution, exchangeable ions).

Despite the multiple causes of EC variability, bulk soil EC measurements have been related to individual factors that limit soil use and productivity such as salinity (De Jong et al., 1979; Rhoades and Corwin, 1981), clay content at a depth of 15-m in New Wales, Australia ($r^2 = 0.78$; Williams and Hoey, 1987), depth of sand deposition along the Missouri River ($r^2 = 0.73$-0.94; Kitchen et al, 1996), depth to claypan in Missouri ($r^2 = 0.73$; Doolittle et al., 1994), and soil moisture content ($r^2 = 0.96$; Kachanoski et al., 1988).

If soil EC maps have utility in production agriculture, 1) EC must be spatially structured, 2) spatial patterns must have temporal stability, and 3) EC must be related to factors of agronomic importance. The objective of this research was to determine the nature and causes of EC variability for several fields in Kentucky. Geostatistical analyses were conducted to examine the spatial and temporal variability of EC variability. Transect studies were conducted to determine the causes of EC variability.

Methods

Site Description

This research was conducted in a field in Hardin Co, KY (Field 1), and three fields in Shelby Co, KY (Field 2, Field 3, and Field 4). Field 1 consists of the Vertrees series (Fine, mixed, mesic Typic Paleudalfs), which formed in residuum from limestone, the Nolin series (Fine-silty, mixed, mesic Dystric Fluventic Eutrochrepts), which formed in mixed alluvium, and the Crider series (Fine-silty, mixed, active, mesic Typic Hapludalfs), which formed in loess over limestone residuum.

Field 2 consists of the Nicholson series (Fine-silty, mixed, mesic Typic Fragiudalfs) and the Lowell series (Fine, mixed mesic Typic

Hapludalfs), both of which formed in residuum from limestone and shale and have a loess cap. Field 3 has both the Nicholson series and the Shelbyville series (Fine-silty, mixed mesic Mollic Hapludalfs), which formed in loess over residuum from limestone. Field 4 contains the Cynthiana series (clayey, mixed, mesic Lithic Hapludalfs) and Faywood series (fine, mixed, active, mesic Typic Hapludalfs), both which are shallow to bedrock and formed in residuum from limestone.

Soil EC Data Collection

A Veris® 3100 Soil EC Mapping System was used to measure soil EC. The sensor consists of six coulters, two of which introduce an electrical potential into the soil. The remaining four coulters are spaced to measure EC over two approximate depths, 0-30.5-cm ($EC_{30.5}$) and 0-91.5-cm ($EC_{91.5}$). When used in conjunction with a DGPS receiver, EC data can be geo-referenced to create a map.

Transect EC Measurements

Transects were selected from Field 1, Field 3, and Field 4. At selected points on each transect, several measurements were taken in addition to EC. Volumetric water content to a depth of 12-cm was determined with the HydroSense™ (Decagon Devices, Inc, Pullman, Washington), which uses transmission line oscillation.

Depth to a clay increase was assessed using the "texture by feel" method. Penetrometer resistance was measured on all Field 1 transects. Depth measurements to fragipan (Field 3) and to bedrock (Field 4) where measured only in the fields where these factors affected crop production.

Soil EC values for each point on the transect were determined by driving directly over the sampling point, stopping, and recording the $EC_{30.5}$ and $EC_{91.5}$ values. In addition, the transect EC values in Field 3 were measured on two consecutive days.

Whole Field EC Measurements

Whole field EC data were collected for a location in Hardin Co. (Field 1) on three dates: Oct. 8, 1999, March 7, 2000, and May 5, 2000. A second location in Shelby Co. (Field 2) was measured once on July 7, 1999. The fields were traversed with approximately 7.5-m between passes on each date.

Field 3 was traversed in the north-south direction and the east-west direction. Data were recorded every second, and groundspeed was maintained at approximately 10.5 km hr^{-1}.

Introduction

Soil samples were collected for the top 15-cm using a 30.5-m grid pattern. Soil analyses, including pH, buffer pH, organic matter content, and Mehlich III extractable P, K, Ca, Mg, and Zn, were performed by the Department of Regulatory Services, at the University of Kentucky. At each sample site, all EC values falling within a 4.6-m radius were averaged and related to $EC_{30.5}$ and $EC_{91.5}$ using simple linear regression.

Results and Discussion

Nature of the Variability

Day-to-day variability was greater for $EC_{30.5}$ than for $EC_{91.5}$. While collecting EC data, extreme values were encountered on occasion, as can be seen by the high value for May 4th $EC_{91.5}$. The small-scale temporal variability also gives an indication of measurement error. The relative nugget variances for both depths also give an indication of measurement error ($EC_{30.5}$, 46%; $EC_{91.5}$, 27%) and were large.

The larger scale temporal variability reflects changes in EC associated with different environmental conditions. While the EC values were substantially lower during the drought of 1999 (October 8th, 1999) than in the spring of 2000, the general spatial patterns in EC were similar across all three dates.

Both $EC_{30.5}$ and $EC_{91.5}$ tended to be higher on the Vertrees series, which is located in the lower right, lower left, and upper right regions of Field 1. The values tended to be lower for the Nolin series, which are mainly located in the depressions.

The remainder of the field was the Crider series. Soil EC was a good indicator of soil type for this field. The Vertrees soils have a red, Bt horizon near the surface, which increases conductivity. The EC values for the Vertrees series were not as high on October 8th, 1999 (during the drought) as on the later two dates.

This suggests that soil moisture enhances the conductivity of clay. This did not, however, greatly change the overall appearance of the maps on the different dates because each was based on an equal number of observations in each category rather than equal sizes of mapping intervals. Date of the measurement did, however, change the spatial statistics of the data.

Anisotropic behavior was not apparent on October 8th, 1999, but it was on the latter two dates. The dark diagonal lines indicate the direction of the anisotropic axis with of minimum spatial variability (here the northwest-southeast direction). Orthogonal to this direction indicates the direction with the maximum spatial variability.

Anisotropy was present in this field because the Vertrees soil, which occurred in southwest, northeast, and southeast regions of the field, had very large EC values on the latter two dates when soil moisture was greater. Therefore, variability was much greater in the southwest-northeast direction. In all cases, the anisotropy does not seem to be a great issue within the first 50-m. Anisotropy was less important with $EC_{91.5}$. This may be because at greater depths, the soil was more uniform in clay and moisture content. The impact of anisotropy in this field would depend upon the depth of interest, whether the data would be used for interpolation, and the sampling interval.

Chapter 2

Soil Testing : How to Test Soil pH

Soil testing is of immense importance, since all the plants show different pH preferences. The pH of the soil directly affects the quality of nutrients that the soil offers. You can do the soil testing of your garden soil with the help of test kits. Read on to know more about how to test soil pH.

Soil pH can influence many factors such as availability of nutrients in the soil, soil life and the susceptibility of plants to various diseases. 'Soil life' is the term used to define various micro-organisms that are responsible for decomposing complex compounds into simpler organic forms, thereby enriching the soil. Some micro-organisms thrive in alkaline soil, while others prefer acidic soil. Soil testing helps you to determine the pH of the soil so that you can make amendments to improve it.

How to Test Soil pH

Soil testing can be done by taking samples of soil from different areas of your garden and mixing them together. This helps to determine the average pH level of the soil. While taking samples, make sure you don't take them immediately after fertilizers are applied or when the soil is wet.

Soil pH Testing With Test Kits

You can get the pH test of your soil done from any garden centre or you can even do it yourself. For that, you'll need to buy a pH test kit which is easily available at the nurseries or hardware stores. The testing components that come with these kits include a test tube, testing solution and a colour chart. For performing a soil test, put the sample of your soil in the test tube and add a few drops of the test solution. Next, shake it well and keep it aside for an hour or so for settling. The nature of your soil makes the solution in the tube to change its colour. The new colour indicates the level of pH of the soil. Compare the colour of the sample in the test tube with the colours on the colour chart. The one that matches with the colour of your sample gives you the pH of your soil. Some kits also come with booklets that help you to interpret your result.

Soil pH Testing Using Other Methods

If you cannot arrange for the testing kit, you can even test the pH of your soil at home using some commonly available ingredients and tools.

Method 1: Scoop some soil from your garden and put it in a container, then add half a cup of vinegar in it. If the mixture fizzes, it means your soil is alkaline. If it shows no reaction, add half a cup of water in freshly scooped soil. Then mix some baking soda in it and see if the mixture fizzes. If it does, it means your soil is highly acidic.

Method 2: Dig a hole 2 - 3 inches deep in the ground and fill it with distilled water to form a muddy pool. Get the test probe and wipe it with tissue paper to make sure it is clean. Insert the probe in the hole and take the reading after 1 minute. The test probe may not cover the entire pH range of 1 - 14, however, it indicates if it is below 7 or above 7. If the scale is below 7, the soil is acidic, otherwise it is alkaline. Exact 7 scale indicates neutral soil.

Method 3: Collect one scoop of soil from different areas in your garden and mix it well. Spread it on a newspaper and allow it to dry. Then take half a cup of soil in a jar and fill the jar with distilled water. Allow the mixture to set until the soil collects at the bottom of the jar. Take litmus paper and insert it in water. If it turns red, the soil is acidic; blue litmus paper indicates that the soil is alkaline.

The degree of acidity or alkalinity can be measured only with pH metre or the colour chart that comes with the test kit. Exact pH level helps you to amend the soil. If the soil is acidic, amend it by adding wood ash or lime, while alkaline soil can be amended by adding sulfur.

Lowering Soil pH - How to Lower Soil pH Level

Plants are conditioned to thrive in garden soil, with the particular pH level that's best for that plant type. Some plants need a more acidic soil to grow and flourish in. Therefore in areas with alkaline soils, there is a need for lowering the soil pH levels. Take a look at these simple ways to lower soil pH levels.

Soil Testing : How to Test Soil pH

For generations, farmers have added limestone or loads of marl to 'make the land sweet and the keep the land up'. Most of us have probably heard these words from ours grandfathers, who believed that lime would make the soil tilth and easy to plow, leading to better crops. Obviously your grandfather did not know anything about soil pH so to say, but that it is essentially what determines whether your garden plants will thrive or not. This leaves us wondering, what exactly is soil pH? In a nutshell, the pH value is a measure of the acidity of the soil and is based on a set of numbers from 1 to 10, that are universally recognised. It is a measure of the concentration of hydrogen ions in the soil solution. While the number 7 is the measure of the soil that is neutral, numbers above 7 indicate an alkaline soil and numbers below 7 indicate an acid soil.

The Need to Lower Soil pH

Most plants grow best where the soil is slightly acidic, in the range of pH 6 to 7. While a few plants, such as azaleas, gardenias and blueberries, grow best at lower pH levels, others such as centipede turf, camellias and potatoes, grow well in a wide range of pH conditions, but seem to flourish in more acidic soils. The other reason for lowering soil pH, is its effect on the nutrient availability for plants, as soils with a pH of over 7.8, have a prevalence of iron, zinc, and phosphorus deficiencies. The high salt levels can lead to yellowing and poor growth of the plants.

The reason for high soil pH can be deemed to the arid climates, with the rainfall not leaching the calcium and other basic materials out of the soil, like the Black Belt prairie region of central Alabama. Sometimes the high pH can be the result of gardeners inadvertently adding more lime to the soil, than needed, without taking a soil test.

Lowering Soil pH: How to Lower Soil pH Level?

The first step in lowering soil pH, is to test the soil using a soil testing kit. There are two basic types of soil testing kits, available. While one is a capsule that will change the colour of the soil & water mixture, that is then viewed against a colour coded chart, the other is a fully reusable probe, with a simple-to-read metre at the top. The next step includes the challenging and slow process of lowering the soil pH, using organic or inorganic soil amendments. Based on the pH, lime content, soil texture, and mineral and nutrient content, you can use any of the following methods to lower the soil pH.

- In the majority of the cases, soil pH can be lowered, simply by using fertilizers containing ammonium, like ammonium sulfate and sulfur coated urea.
- You can also amend the soil by adding sulfur, that is available in the two forms of dusting sulfur and aluminium sulfate. While dusting sulfur might take several months to correct the soil pH, aluminium sulfate has a more immediate effect. A costlier but effective way to lower soil pH, is by using iron sulfate. While adding sulfur is mixed with the soil, it is important that the soil is moist, aerated and warm, to enable the rapid growth of bacteria.
- For those who prefer the more organic method, compost acts as a buffer to protect plants from unbalanced soil pH. You can use decayed vegetable matter, compost, stable manure and straw, etc. to increase the acidity of the soil. This method allows the pH to be slowly lowered over time, while increasing microbial life and improving the structure of your soil.

It is important that before acidifying the soil, a gardener ascertains the reason as to why pH levels are high and the soil's type. So, sandy soil would require less amendment than clay soil. Ensure that while applying the soil amendments, the correct amount of the product is added.

Soils that are over acidified should be limed, to neutralise soil pH to the desired soil pH level. Once the soil pH has been acidified to the desired level, it has to be monitored over time with regular sampling and soil analysis.

Soil Types

Soil types are identified with respect to the amount of clay, silt and sand particles present in the particular soil sample. Different types of soil and their basic properties are highlighted in this article.

Soil is defined as the topmost material layer of the earth's surface, in which plants grow and spread out their roots. It is made up of organic material (decayed plants, animals), inorganic matter (rock, mineral), air and moisture. From a layman's point of view, soil seems to be the same everywhere, except for extra dryness or soggy state. But in the true sense, soil types are many, each of which varies with respect to particle size, water holding capacity, resistance to erosion and many such aspects.

Types of Soil: Explained

In soil science, soil types and their identifying properties are studied in detail. There are three kinds of soil particles, namely, sand, silt and clay.

The percentage composition of these particles helps in determining the soil type of a particular region. To be a good gardener, you need to identify the types of garden soil and search for plants that suit the soil condition. Even if you do not have a yard, and are growing vegetables and flowers in pots, then also you need to know the types of potting soil. Listed below are different types of soil along with their specific characteristics.

Clayey Soil Type

Clay is the smallest soil particle, characterised with increased water holding and nutrient binding properties. Since it has the ability to hold water, clayey soil is usually heavy, sticky and the rate of saturation is high. True clayey soil is not preferred for plantation purpose, as there is lack of soil aeration. This type of soil can be used after mixing with peat and sand particles.

Silty Soil Type

Another in the list of fertile and heavy soil types is silty soil. Available mostly in estuaries, it is composed of small particles (but larger than clay) and has the ability to hold water. Similar to clay, silty soil turns sticky after saturation with water. As per soil testing results, its texture is soft and smooth, and water draining is better than clay soil.

Loamy Soil Type

Loam is one of the ideal soil types for plant growing purposes. It comprises proportionate amounts of sand, silt and clay in the ratio 40:40:20. Generally, loam soil is fertile (unlike sandy soil) and has no water drainage problems like clayey soil and silty soil. In short, it is fertile and well-drained soil, excellent for cultivation.

Sandy Soil Type

As the name goes, sandy soil is basically made up of sand particles. It is lightweight and can be dug up very easily. The individual grains are large sized, thus added for increased aeration and easy draining. Nevertheless, sandy soil alone is not good for planting purpose, as it does not retain water, nutrient and fertilizers.

Peat Soil Type

Peat soil is loaded with organic materials (decaying remains of plants and animals). In comparison to other soil types, peat soil has the highest organic matter. Considering this, it is understandable that this is an acidic soil type (low pH range). Peat soil with moderately high pH and good water draining ability is good for plantation.

Chalky Soil Type

In contrast to peat soil, the chalky soil type is alkaline in nature (high pH range) and prone to dryness. Also known as basic soil, it holds very less moisture, enabling water to drain off very easily. Despite the fact that chalky soil contains essential plant nutrients, they are not available to plants due to increased alkalinity.

In case, the plant variety that you are interested in growing is not suitable with the garden or potting soil types, you can mix it with different soil particles until you get the desired characteristics. This versatility has made gardening an interesting hobby, as soil condition plays a crucial role in determining the bloom size of flowering cultivars and yield of crops. After all, soil is the medium, from which plants derive their nutrients for growth and development.

Acidic Soil

What is acidic soil? How to make soil acidic? Which are the plants that like acidic soil? Scroll down for answers to these questions.

Soil Testing : How to Test Soil pH

Every plant requires specific conditions to grow. Be it the temperature range, amount of sunlight, amount of water, soil nutrients or the pH of the soil! They all have a set of requirements for proper growth. In this article, we shall discuss acidic soil and the plants that grow in acidic soil.

Soil Acidity

Some plants require a fairly acidic soil to grow, while others grow well in alkaline soils. Soil pH is actually the measure of its acidity. Soil having a pH of anything below 7 is considered as acidic soil. Soil acidity is one of the most important factors that decides the growth and overall health of the plant. The pH of the soil greatly affects the availability of nutrients in the soil. A number of diseases that affect plants also tend to thrive in soil that has a particular pH. This is the reason why determining the pH of the soil is an important part of gardening. Read more on soil pH test.

How to Make Acidic Soil?

Certain plants like blue hydrangea or azalea thrive only in acidic soils and hence, knowing how to make acidic soil becomes very important for gardeners who intend to grow such acid loving plants. The pH of the soil can be determined with the help of kits available in your local nursery or you could get it checked by sending a soil sample to the lab.

If you find that the soil is alkaline you will have to lower its pH, to make it acidic. You can do this by adding certain elements to the soil. This is how you could go about it:

- Adding sphagnum peat to the soil is the easiest way to lower the pH of the soil. Just add a pinch of peat moss to the soil around the plants. You could also add it during planting.
- Another inexpensive way of increasing the acidity of soil is by adding sulfur to the soil. However, this may take a few months to make the soil acidic as sulfur is one of the slowest acting options.
- Watering your plants with two tablespoons of vinegar mixed with a gallon of water, twice a day can be an easy solution to fix soil alkalinity. However, this is a temporary method and will last only till the vinegar is absorbed and depleted.
- Putting a layer of sawdust or wood chips on the soil around plants can also help in lowering soil pH.

- There are several organic fertilizers available that help in increasing the acidity of soil. Fertilizers containing ammonium nitrate, ammonium sulfate or sulfur coated urea serve as a good choice for lowering the pH of the soil.

Plants that Grow in Acidic Soil

Some examples of plants that like acidic soil include:
- Heather - *Calluna vulgaris*
- Azalea - *Rhododendron kaempferi*
- Wintergreen - *Gaultheria procumbens*
- Bayberry - *Myrica pensylvanica*
- Daphne - *Daphne cneorum*
- Mountain Laurel - *Kalmia latifolia*
- Silverbell - *Halesia carolina*
- Holly - *llex opaca*
- Oak - *Quercus palustris*
- Sweetbay - *Magnolia virginiana*
- Summersweet - *Clethra alnifolia*
- Fothergilla - *Fothergilla major*
- Blueberry - *Vaccinium corymbosum*
- Bear's breech - *Acanthus mollis*
- Kiwifruit - *Actinidia deliciosa*
- Red buckeye - *Aesculus pavia*
- Wild sarsaparilla - *Aralia nudicaulis*
- Sweet fern - *Comptonia peregrina*
- Winter hazel - *Corylopsis pauciflora*
- Chinese cucumber - *Cucumis melo*
- Star magnolia - *Magnolia stellata*
- Royal fern - *Osmunda regalis*.

This was all about acidic soil and plants that thrive in acidic soil. Almost all the evergreens prefer acidic soil.

Rhododendrons, hydrangeas, azaleas, lilacs and most varieties of roses are some more examples of acidic soil plants. Hope you found this information useful!

Soil Horizon Layers

Soil is made of a number of distinct, horizontal layers placed one above the other. Each of these distinct layers is known as a soil horizon. Read on to know more about what is soil horizon and its various layers...

Along with wind and water, soil is the major natural resource that supports life on earth. It is so important that a whole branch of study called soil science has developed around it. It is made of three components - minerals, organic matter and the living organisms that live in its upper layers. Soil is formed by the weathering action of natural elements like wind, water, glaciers and change in temperature. Soil can be formed from the rocks lying below or from rocks present somewhere far away. These agents of weathering progressively break rocks into finer grains that are laid in layers to form the soil.

What is Soil Horizon?

Soil is made of distinct layers that lie one above the other, parallel to the soil surface. Each distinct layer is called a soil horizon. A vertical cross-section of a soil known as the soil profile reveals the various horizons of the soil. Each soil horizon is the result of a number of geological, chemical and biological processes that have been taking place for over thousands of years. Hence, the soil horizons are best formed and delineated from each other in older soils. The various soil horizons are identified on the basis of physical features, mainly their colour, texture and particle size. Though the soil composition varies from place to place, most soils conform to a general pattern consisting of six horizons.

Soil Horizon Layers

The soil contains the following physically distinct horizons from top to bottom:

O Horizon

The letter 'O' stands for organic. As the name suggests, this horizon is rich in organic material of plant and animal origin. These materials are generally in various stages of decomposition. This decomposed organic material is called the humus that gives this horizon its characteristic dark colour.

A Horizon

This is also known as the 'topsoil', and it is the topmost layer of the mineral soil. However, as it lies just below the O horizon, this layer also has some amount of humus in it. Hence, it is darker in colour than the layers lying below it. This layer is also known as the 'biomantle' as it is the A horizon in which most of the biological activities take place. Soil organisms like earthworms, fungi and bacteria are mainly concentrated in this layer. The soil particles in this region are smallest and finest as compared to the lower horizons of the soil.

E Horizon

This layer lies below the A horizon and above the B horizon. It is light in colour and contains mainly sand and silt. It is poor in mineral and clay content as these are lost to the lower layer by the process of leaching. Hence, this horizon is also called the layer of eluviation (leaching). The soil particles of this layer are larger in size than those in the A horizon but smaller than those in the underlying B horizon.

B Horizon

This is referred to as the 'subsoil'. This lies just below the E horizon and is rich in clay and minerals like iron or aluminium. Though this layer has a higher mineral content than the topsoil, some organic material may reach this layer from the layers above by the process of leaching. Plant roots may reach this layer. However, the B horizon is reddish or brownish due to the oxides of iron and clay.

C Horizon

This layer is also known as regolith. The C horizon is mainly made of large rocks or lumps of partially broken bedrock. This layer is least affected by weathering as it lies deep within the soil and is inaccessible to the soil-forming agents. Hence, the rocks in this layer have changed very little since their origin. Plant roots do not reach so deep down to this layer. The C horizon is typically devoid of organic matter.

R Horizon

This is the bedrock. It is the deepest soil horizon in the soil profile. Unlike the above layers, this horizon does not consist of rocks or even boulders. It is made of continuous mass of bedrock. Digging through this layer is very difficult.

Study of the soil horizon is the firs step towards soil taxonomy. These layers help us to understand the geological events of the past, and the properties of the soil in general.

Role of the Extension Service in Soil Testing

The actual analysis of the sample and the making out of fertilizer recommendation is only part of the soil testing service. To a large measure, the efficiency of this service depends upon the care and effort put froth by extension workers and farmers in the collection and dispatch of samples to the laboratory. Its effectiveness also depends upon the proper follow-through of the fertilizer recommendations, including the establishment of result demonstrations on farmer's fields to induce the farmers to follow the fertilizer recommendations. In this work the staff of the extension service play the most important role, since they are the people directly in contact with the farmers or this reason, the soil chemist in charge of the laboratory must give periodic and through training to the extension staff on these subjects.

Collection of Samples

A useful soil testing service starts with the collection of representative soil samples. A fertilizer recommendation made after analyzing the soil can only as good as the sample on which it is based. Actually the one to ten grams of soil used for each chemical analysis should represent as accurately as possible the entire surface six inches of soil, weighing about 2 million pounds per acre. The importance of taking a representative composite sample is, therefore, self-evident. One field can be treated as a single sampling unit only if it is relatively uniform and does not exceed approximately five acres. Variations in slope, colour, texture, management, and cropping pattern should be taken into account and separate composite soil sample adequately representing the field, small portions of surface soil should be collected to depth of six inches from at least ten well-distributed spots in the field, mixed well, and about ½ kg of representative sample sent to laboratory.

Proper sampling tools are essential for collection of good soil samples. For a soft. Moist soil, the soil tube, phowda (spade), or khurpi (trowel) are usually quite satisfactory.

For harder soils, a screw type auger, or an adze might be more convenient. Post hole augers are convenient for sampling excessively wet areas like paddy fields. An extension worker whose duties include collection of soil samples should be supplied with at least a few of these tools, and also a plastic bucket. The phowda, khurpi and adze are very common implements available in most hardware shops and so there should be no difficulty in procuring these implements.

The farmers should be given help in filling out the soil sample information sheet with an ex-plantation of any items not understood. It should be remembered that the information sheet is very vital part of procedures that go to make a good soil test recommendation. This sheet must supply all of the background information that, in combination with the results of the analysis, makes possible an accurate fertilizer recommendation for a certain crop, for that particular field.

Factors such as crop variety, slope of land, irrigation and drainage facilities, and pervious cropping seasons affect the amounts of fertilizer to be applied to particular crop. Any peculiarities noted in the soil or in the vigor or the crop would be very valuable information on the soil sample information sheet as a basis for making an adequate fertilizer recommendation. In the absence of this information, the soil chemist must base his recommendation upon the soil test values alone, and more often than not the farmer will receive an adequate fertilizer recommendation.

Soil Analysis: A key to Soil Nutrient Mangement

High yields of top-quality crops require an abundant supply of 16 essential nutrient elements. In addition to providing a place for crops to grow, soil is the source for most of the essential nutrients required by the crop. Our soil resource can be compared to a bank where continued withdrawal without repayment cannot continue indefinitely. As nutrients are removed by one crop and not replaced for subsequent crop production, yields will decrease accordingly. Accurate accounting of nutrient removal and replacement, crop production statistics, and soil analysis results will help the producer manage fertilizer applications.

A soil analysis is used to determine the level of nutrients found in a soil sample. As such, it can only be as accurate as the sample taken in a particular field. The results of a soil analysis provide the agricultural producer with an estimate of the amount of fertilizer nutrients needed to supplement those in the soil. Applying the appropriate type and amount of needed fertilizer will give the agricultural a more reasonable chance to obtain the desired crop yield.

Objectives of Soil Analysis

- To provide an index of nutrient availability or supply in a given soil. The soil extract is designed to evaluate a portion of the nutrients from the same "pool" used by the plant.
- To predict the probability of obtaining a profitable response to fertilizer application. Low analysis soils may not always respond to fertilizer applications due to other limiting factors. However, the probability of a response is greater than on a high analysis soil.
- To provide a basis for fertilizer recommendations for a given crop.
- To evaluate the fertility status of the soil and plan a nutrient management program.

Chemical analysis of plant composition indicates chemicals or elements present in a crop at maturity or when it is harvested. For example, 1,250 lb of lint cotton contains approximately 125 lb of nitrogen (N), 20 lb of phosphorus (P), and 75 lb of potassium (K).

The essential question in fertilization is, "How much nutrient must be added to the soil as fertilizer for a given amount to be taken up by the growing plant?" The crop utilizes only a portion of the available nutrients in the soil. This means that more nutrients must be present than are removed by the crop. The amount added varies according to the level already present in the soil and the crop's need for the nutrient involved. The soil analysis is the starting point, since it measures the level or content presently in the soil.

The soil analysis along with the information provided in the information sheet, is interpreted and reported in terms of the nutrients needed to supplement those in the soil. With this information, producers can add sufficient nutrients for the correct balance to obtain high yields.

Limiting Factors

Crop yields are determined by a variety of factors including crop variety selection, available moisture, soil fertility, crop adaptation to the area, and the presence of diseases, insects, and weeds. The soil analysis and its interpretation deal only with the fertility level (plant nutrients) of the soil. Recommended fertilizer will provide sufficient nutrients for the best possible yields. Other factors of production or management may still cause low yields, even though nutrients are adequate.

Carryover

If yields are only partial in relation to a large amount of fertilizer applied, many of the nutrients are carried over for use by the next crop. It is this carryover, or residual effect, from one year to the next that makes heavy fertilizer applications practical in the face of other limits to yield.

Yields to Expect

A certain fertilizer application cannot be expected to produce a specific yield such as two bales of cotton or nine tons of hay. It is more realistic to assume that a balanced fertilizer program assures that nutrients are not the limiting factor in yields obtained. Research has shown that producers who use a balanced fertilizer program obtain consistently better yields than those who don't.

The Soil Analysis Report

After the soil is analyzed, fertility recommendations are made based on amounts of actual nutrients in the soil, not on the amount of any particular fertilizer or mixture. For example, if 100 lb of N were recommended, that amount could be supplied by approximately 300 lb of ammonium nitrate (33%N), 220 lb of urea (45%N), or 120 lb of anhydrous ammonia (82%N). Likewise, a recommendation of 60 lb of P_2O_5 per acre could be added as 133 lb of 45% triple superphosphate.

Fertilizer Labeling

Nitrogen is expressed on the elemental basis as "total nitrogen" (N). Phosphorus is expressed on the oxide basis as "available phosphoric acid" (P_2O_5). Potassium is expressed as "soluble potash" or potassium oxide (K20). In reality, there is no P205 or K20 in fertilizers. Phosphorus exists most commonly as monocalcium phosphate, but also occurs as other calcium or ammonium phosphates. Potassium is ordinarily in the form of potassium chloride or sulfate. Furthermore, P205 and K20 are not absorbed by plants. Plant roots absorb most of their phosphorus in the form of orthophosphate ions, H2P04-, and most of their potassium as potassium ions, K+. For these reasons, the elemental expression (N-P-K) is used in all of the recent research publications. Conversions from one form of P and K to another can be made using the following formulas.

%P=%P205 x0.437   %K=%K20x0.826
%P205=%Px2.29   %K20=%Kx1.21

Interpretation of Soil pH

In England & Wales target pH (measured in water) is 6.5 for arable crops and 6.0 for grassland. Optimum range in Scotland where Scottish Agricultural College (SAC) measure pH in calcium chloride is 6.0-6.5 for arable crops and 5.7-6.2 for grassland.

In-field measurements using pH indicator on some soils where free chalk or lime particles exist may give lower values than laboratory results for the same field. This is because grinding the soil for laboratory analysis pulverises any chalk/lime particles and the pH as measured is increased. Acidity below pH 6.0 will reduce the availability of nutrients, especially P.

Soil analysis may overstate the availability of K on some alkaline soils where pH is over 7.0 and responses to added potash may occur at higher soil K levels than would be expected.

Availability of trace elements is radically affected by pH and the need for trace elements should only be assessed after any required amendment of acidity has been undertaken and has had time to take effect.

Interpretation of Soil P, K & Mg

Soil analysis provides an estimate of available P, K and Mg concentrations in soil to sampling depth-in practice this is equivalent to plough or cultivation depth because of the distribution of nutrients when the land is worked. Response experiments with different crop groups have provided the relationship between crop yield and soil nutrient concentration. Normally, yields increase with increasing nutrient concentration to a maximum, beyond which there is no further benefit from additional nutrient. Below this value, which will vary with crop species, there is a yield penalty. Whilst soil analysis is not a precise guide, the lower the value the greater the risk of poor performance.

To aid interpretation of the different concentrations of individual nutrients, Index or descriptive scales are used. These scales provide a general indication of the likely crop response and therefore a guide to the need for additional nutrient supplementation, as shown in the table.

Crop response and soil analysis

Defra Index grass	SAC description	Yield response to added nutrient by	
		vegetable crops	arable crops &
0	Very low	highly likely	highly likely
1	Low	highly likely	probable
2	Moderate	likely	unlikely
3	High	possible	nil
4	High	unlikely	nil
5	High	nil	nil

Soil P, K & Mg Concentrations (mg/l) and Defra Index Scale

Note that the index is split in half for potassium only and described as 2- (or lower index 2) and 2+ (or upper index 2). In the past, index 2 was not divided in half for potassium but some soil reports used + and -signs to denote the extreme top and bottom 10% of each band; laboratories should no longer be using this convention.

Besides providing a basis to decide fertiliser quantities, soil analysis should also be used to monitor changes in fertility especially where there are uncertainties in the amounts of nutrient removed (e.g. with forage crops) and in the amounts of nutrients applied (e.g. with manures and slurries).

For this purpose it is desirable to use the mg/l values not the index. However differences of less than 5 mg/l Olsen P, 25 mg/l K and 10 mg/l Mg should be ignored unless part of a sustained trend. Where accurate nutrient balance information is used in conjunction with regular soil analysis, it is important to recognise the possibilities of variation as discussed above.

Soil P, K and Mg Concentrations (mg/l) and SAC Descriptive Scale

The Scottish Agricultural College laboratory uses different extractants to those used in England and Wales and a descriptive rather than a numeric scale.

Description	Phosphorus	Potassium	Magnesium
	Modified Morgans extraction		
Very low	0-10	0-40	0-20
Low	10-25	40-75	20-60
Moderate	26-75	76-200	61-200
High	76-200	201-1000	201-1000
Excessively High	201-	1001-	1001-

Relationship between Defra and SAC Scales

Defra Index	SAC description
0	Very low
1	Low
2	Moderate
3-7	High
8-9	Excessively high

Principles of P, K & Mg Manuring

The principle of manuring is to maintain plant-available soil nutrient levels within a target range depending upon crop rotation and soil type, by replacing nutrients removed. Soil analysis shows the status of soil nutrients relative to the target values and allows changes as a result of husbandry to be monitored. Where soils are below the target level, nutrient applications should provide more than is removed by the crop to ensure full yield response and to improve nutrient levels. Nutrient applications for soils above the target range may be reduced or omitted until the soil reserve approaches the target value. Additional nutrients may be applied before very responsive crops such as potatoes with the surplus balance being allowed for in subsequent less responsive crops. The overall nutrient balance in the rotation should be estimated and checked by regular soil analysis.

Sandy Soils

On true sand textured soils (not loamy sands or sandy loams) it is not practical to attempt to maintain soil K at 2- and the target K level should be adjusted to 100-120 mg/l (index 1+).

Chalk and Limestone Soils

At very high pH, potash additions are more likely to be converted to slowly available reserves in the soil and it will be more difficult to raise the soil K index. However soil K targets and replacement principles for application remain the same as for other soils.

Soil pH Levels

The significance of soil pH in gardening is discussed in the following article. The information about how to check soil pH levels and the pH requirements for different vegetables.

The measure of soil pH is considered to be important in determining its health and in deciding the plants which would be best grown in it. Let us understand the different concepts related to soil pH in the following paragraphs.

How to Check Soil pH Levels?

The measure of acidity or alkalinity of soil on a scale of 14 is referred to as soil pH. In order to test the soil pH one would require a trowel/spade to dig the soil, distilled water, and a tester. The hole that one digs needs to be 2-4 inches deep. Twigs or other foreign bodies, if any, should be cleared off and the distilled water should then be poured to create a muddy pool. The tester should be calibrated properly, cleaned, and then inserted into the soil. It should be held in the mud for about 1 minute in order to measure the pH. If the reading indicates 7, the soil is neutral. The reading above 7 indicates that it is an alkaline soil, while a reading is below 7 indicated soil that is said to be acidic in nature. This is one of the most common and accepted ways of testing the pH value of soil.

Reason for Testing Soil pH

The pH of soil can also affect the availability of nutrients to plants. One of the reasons behind giving importance to soil pH is that occurrence of some of the diseases is dependent on the pH of the soil. Also, different vegetables and plants require different levels of soil pH to grow. Gardening with improper levels of soil pH can result in a poor yield of crops. Given below are the pH levels required for grass and vegetable, along with the explanation of some of the common terms and equipment used in soil testing.

Soil pH Levels for Grass

A slightly acidic soil is required for proper growth of grass. Soil pH between 6.5 and 7.0 is considered to be suitable for grass. Ground limestone should be added to the soil if the pH is less than 6.5, while sulfur needs to be mixed in an alkaline soil with pH greater than 7.5. Read more on lawn care.

Soil pH Levels for Vegetables

Vegetables need a slightly acidic soil for healthy growth. For most of the vegetables, the pH requirement is between 6 and 7.5. The soil should thus, be tested for the pH in order to grow healthy vegetables. The vegetables, along with a proper pH level, also need well drained and friable soil.

Soil pH Level Tester

The pH tester provides the farmers/gardeners with the readings about the acidity and alkalinity of soil. It is possible to determine which plant to grow in the soil after knowing the soil pH.

Soil pH Level Chart

The pH level chart for different vegetables given below would help in determining the pH levels for gardening. The pH requirements of some of the commonly grown vegetables are provided in the following chart.

Vegetables	Required pH
Cabbage	6.0 - 7.5
Cauliflower	5.5 - 7.5
Cucumber	5.5 - 7.5
Lettuce	6.1 - 7.0
Onion	6.0 - 7.0
Potato	4.5 - 6.0
Pumpkin	5.5 - 7.5
Spinach	6.0 - 7.5
Tomato	5.5 - 7.5
Beans	6.1 - 7.5

Lowering Soil pH

It is necessary to lower soil pH in order to grow plants which prefer acidic soils. One should test the soil and the check the pH. Fertilizers like sulfur coated urea and ammonium sulfate are used in lowering soil pH. Aluminium sulfate or simply sulfur can be dusted to lower the soil pH. This is especially recommended for soils which have a high content of free lime. Read more on soil science.

The importance of soil pH levels in gardening/farming and the relevant information is discussed in the article above. The chart which provides data about pH requirements of vegetables would also prove to helpful. The information about soil testing too is covered in short. Thus, a comprehensive account of the concept of soil pH is presented above.

While most gardeners and farmers would most likely know all about soil and the pH levels required for different plants, it is less likely that the average individual will know all the details. For those planning on growing vegetables in their garden, or simply for those who want a picture perfect lawn, the above article about soil pH levels will definitely come in helpful.

Soil pH Test

Soil pH test plays a significant role in checking the levels of soil pH levels. Knowing the correct soil pH is essential for the growth of the plants.

Soil Testing : How to Test Soil pH

The pH value of the soil measures the acidity and alkalinity of the soil. The soil pH test helps you to know the amount of nutrients required by the plants and the type of plant you can grow in a particular soil.

Most plants prefer a soil pH of 6.5 to 7.2, as this is a good pH condition for a healthy thriving of a plant. Plants like the rhododendrons love to grow in acidic soil conditions, with the presence of sufficient amount of iron in the soil. Soil pH testing is mainly done to check the presence of major and minor nutrients that the plants require. Checking the soil pH levels is important to ascertain the lack of any nutrient in the soil. Loss of nutrients will prove fatal for the plants. You can add fertilizers in the form of the nutrients to build up a healthy growth for the plants.

Soil pH Test Kits

Soil pH test kits are commonly available in many nurseries and hardware stores. You can get the soil pH checked from any good garden centre. The soil pH kit comprises a test tube, a testing solution and a colour chart. Take a little quantity of the required soil sample and put it inside the test tube, then add few drops of the given test solution in the test tube. Shake the test tube thoroughly and keep it aside for about an hour so that the mixture settles. After one hour, you will notice that there is a change in the colour of the solution. The change in colour shows the pH level of the soil. Now, by the help of the given colour chart, you can compare the solution colour, with the colour on the chart. The colour which matches with the solution colour, gives you the pH value of the soil.

Soil Testing Methods

It is always recommended to test the soil from a local laboratory, as the local people have good knowledge of the local soil. Soil testing is mostly done to check the presence of certain factors, such as availability of the essential nutrients in the soil, testing the moisture content and organic matter in the soil and also, to check the presence of contamination in the soil. The soil pH testing procedure, includes the following methods.

First Method

Collect some soil sample from the garden and pour it in a container. To that, add half cup of vinegar and observe the reaction. If the mixture becomes bubbly or frothy, then the nature of the soil is alkaline. If there is no reaction, again scoop out some soil and add half cup of water to it. Add little amount of baking soda to the soil and water mixture. If still it does not fizz, then the nature of the soil is acidic.

Second Method

Make a muddy pool by digging two inches deep in the ground and cover it with distilled water. You can buy a test probe and clean it with a tissue paper. Introduce the test probe in the hole, and hold it for one minute. Take out the test probe and check the readings, after one minute. The test probe may show you a reading of above 7 or below 7, which can be found out from the markings on the test probe. Acidic soil can be identified with below pH 7 levels, while alkaline soil is indicated with above pH 7 levels. A neutral pH is identified, if the reading is exact 7.

Third Method

Scoop out some soil from various regions of your garden and mix it uniformly. Take a newspaper and disperse the soil, and wait till it completely dries out. Then, take half a cup of the mixed soil in a container and add distilled water to it. Mix it thoroughly and leave the mixture aside for settling down the soil particles. Check the acidity and alkalinity of the soil by dipping a litmus paper in the top water portion of the mixture. A red colour change in the litmus paper indicates acidic soil, while blue colour change represents alkaline soil. Read more on soil science.

Checking the soil pH by soil testing regularly, is the key to good gardening. Soil pH test, if not done properly and in time, can cause diseases in plants and as a result, poor yield of crops. Organic fertilizers and soil amendments are good ways of feeding the soil for a better and healthy thriving of the plants.

Soil Pollution Facts

The effects of soil pollution are becoming increasingly visible with each passing day, crop failure and high levels of water contamination all leading to diseases in humans and animals alike. To place preventive measures in place, lets us first understand a few soil pollution facts.

Life on earth exists in a very delicate balance, where soil, air and water sustain not only human life, but the entire eco-system. Any imbalance in this eco-system due to environmental pollution results in contamination and sets off a chain of disruption that affects all patterns of existence. Soil pollution is a reality today with as severe repercussions as water and air pollution. Soil pollution facts need to be understood, and more importantly controlled.

Soil Pollution Causes

Soil pollution starts with the flawed concept of throwing trash on the side of a road and throwing out your dustbin on the road, and by road I mean dumped in unused yards outside city limits or in open fields. Besides the tons household plastic, industrial dumping of man-made chemicals is also done. As hard as it may be to believe, giant chemical companies dig up large holes, throw the waste in and cover it back up! This sad reality is not just restricted to developing countries, but highly developed and advanced countries as well. Agricultural advancement has also played a part in laying many a green pastures barren. The need for high yield coupled with greed has witnessed incessant and discriminant use of pesticides and chemical fertilizers has stripped the soil of its natural pH balance as well as the soils ability to replenish itself naturally, leading to increased soil erosion. Other causes of soil pollution or contamination include the rupture of underground storage tanks, seepage of contaminated surface water to subsurface strata, chemicals, industrial wastes, oil and fuel dumping. Major soil chemical pollutants are petroleum hydrocarbons, solvents, and other heavy metals. Read more about causes and effects of land pollution and soil pollution causes and effects.

Facts about Soil Pollution

Soil is a non-renewable resource with more potential to degrade. According to the study published by the European Commission in April 2002, among other major threats to soil, decline in organic matter and contamination needs immediate attention. One of the most severe complications of soil pollution facts is that the chemicals from the soil will contaminate the crops grown on them, and also the groundwater that is used for drinking. The same contaminated soil also has the potential to seep into large water bodies and have an effect on the overall eco-system, thus, becoming a major environmental issue. In most countries that have very little control on soil pollutant dumping, or have laws that can be conveniently bent, soil is contaminated with over a hundred odd active pesticides that damage the immune and endocrine systems causing cancer, multiple birth defects and gene mutation, not only in humans but also in animals as well. In U.S. alone, millions of tons of chemical waste is being dumped in the soil and sea, and spewed in air resulting in long-term adverse implications on life in general. Farms are using an excess of nitrogen to increase productivity, and although nitrogen is essential for plant growth, too much of it results in nitrate pollution in the crops, soil and ground water. Read more on facts about land pollution.

Developed and developing countries have now put a major legal framework and clean-up program in place, to deal with soil pollution. One of the important soil pollution facts is that at the end of the day, figures and who is responsible won't matter, what will matter is how long will it take us and the generation to come to clean the mess, that is, if there is anything left to save.

Soil Testing : How to Test Soil pH

Analysis and Collection of Soil Samples

Back in the nineteenth century Edmund Locard developed the theory known as the "Locard's Exchange Principle." The theory in short states that when ever someone comes in contact with another object or person there is a minute exchange of particles that, in theory, can be traced back to a victim or suspect (Block 1999). In a crime context, that evidence can confirm or disprove a hypothesis of involvement between a suspect and a victim. Locard described these particles as dust or dirt but today it is understood to include all soil borne trace evidence. Trace evidence can include blood, hair, fibre, dirt, glass particles and any other minute particles at a crime scene.

It would be nice to think that this forward thinker, Locard, moved beyond his Principle to developed the idea of soil analysis but he didn't. His mentor and friend Hans Gross published articles almost simultaneously with Sir Arthur Conan Doyle, the author of the Sherlock Holmes Fictions, regarding soil analysis (Block, 1999)(Nickel & Fischer, 1999) To Edmund Locard's great disappointment both Arthur Conan Doyle and Hans Gross beat him to the punch on the evaluation of soil as a forensic tool. However, Locard's zeal for the use of trace evidence in forensic investigations led to a life long study of the classification and identification of soil samples (Block, 1999). Since its dawn in the late 1800's the analysis of soil has grown into a multidisciplinary field of forensic study. Modern analysis of soil may involve geologists, entomologists, toxicologists, biologists, botanists and a myriad of other experts.

Soil may be understood as just that: soil. As such it has a value in forensic studies. However, soil may also be understood as a repository for non soil contaminants which can yield valuable information about crimes. In forensic analysis materials can be grouped in several ways and each lab has its own way of subdividing these groups (Chayko & Gulliver 1999). In general soil borne materials can be considered organic or inorganic. Both types are found in soil. Further refining of these classifications where soil is concerned is to break the groups into mineral, biological or synthetic matter.

Soil is considered trace evidence. Soil is made up of disintegrated surface material which can be organic, mineral or synthetic. The ratio of the mineral content compared to other matter in the soil can be very site specific. The ratios of mineral, organic and synthetic matter can vary even with in a few feet (Chayko & Gulliver, 1999). Sandy soils look, feel and behave quite differently from clay soils or peaty soils. By profiling an array of characteristics of each soil, it is somtimes possible to attribute those characteristics to a specific location (Steck, 2004). Soil samples when properly taken can tell an investigator a lot about where a victim or suspect has been. Analysis of soil samples taken from vehicles can also tell an investigator about where a vehicle has been. Analysis of foot wear, clothing and tires can also place a suspect or victim in a particular location.

Collection of Samples

Collection of soil samples will depend on the circumstances of the crime. Indoor scenes will differ markedly from outdoor scenes in the type of evidence that can be recovered and the way in which these samples are collected.

At indoor scenes there may be footprints in soil or in dust. Samples made by footwear should be photographed to scale before being recovered. The particle samples can best be collected using a vacuum method. The samples can be vacuumed with a portable vacuum cleaner equipped with a special attachment.

The attachment has a metal screen on which a filter paper is attached. The area is vacuumed and the filter is removed and labeled with the date, location, time and name of the technician who operated the vacuum. The vacuum must be thoroughly cleaned between samples. Cleaning can be fairly easily done with handhelds where the parts are easy to access. Reference samples from the surrounding area perhaps including flower gardens, points of entry and exit and alibi locations should also be taken.

Information on obtaining these specialized vacuum attachments can be obtained through Sirchie Laboratories Inc. in Youngsville N.C.

In the case of a break and enter or other crime at a home or business it is useful to know how the perpetrator entered the building. For example, if the perpetrator stood in a flower garden outside a window this indicates a stranger. If the perpetrator walked up the front lawn or to the back door this could indicate someone familiar with the property. Possibly the perpetrator knew no one was there being familiar with the owners habits.

The soil on the shoes of a suspect (or left on a carpet or floor) can indicate the direction of travel and the mode of entry. This type of evidence is most useful when a suspect can be immediately identified before the soil is lost from the footwear. Soil evidence from a victim or suspects clothing can indicate an association between victim and suspect. For instance, if a suspect lives in a particular neighborhood with a specific soil profile and this profile matches one found at a crime scene or on a victim, then one must suspect that the suspect left that sample at the scene.

"Double transfer" is even more convincing. If soil profiling to a suspects home territory is found on a victim or in their home and soil from the victims home territory is found on the suspects clothing or footwear the probability of the suspect having been at the crime scene increases dramatically.

The mathematic probability of two matching profiles being found at both scenes by coincidence increases. For example the probability of transfer to one site occurring by coincidence might be 1 chance in 800. When double transfer has occurred the probability of that occurring by coincidence becomes 1 chance in 64,000 (Crocker, 1999).

Inorganic or manufactured matter found in the soil recovered from a suspect's footwear or clothing can be very site specific. Particles of glass, rubber and other industrial products can be used to link a suspect with a particular location. In the not-uncommon case of crimes at industrial locations this can be particularly useful.

How to Collect Samples

As already mention in the indoor scene vacuuming can be used to collect samples. The methods of collection will vary from scene to scene. In a break and enter where house plants are found upset a sample of the soil may be used to link the suspect to a scene. Soil from the garden or yard can also be used to track the suspect's direction of travel and point of entry. Most garden soils are unique. The gardener in charge will have added some favorite materials (sometimes specific to particular plants) in order to improve the garden. The materials tend to be different in every garden. Some gardeners might use compost while others might prefer ammonium pellets, sand, peat or wood chips.

Samples taken from the interior of a vehicle can indicate many things. A soil sample from the gas or brake pedal of a vehicle can link a suspect to a location. Soil from a trunk or backseat can indicate digging. When graves are dug to hide bodies the various levels of the soil are disturbed. If the suspect lays a shovel on the back seat or in the trunk the soil from the lower levels of the grave may be deposited on the seat or in the truck. Soil from tools such as shovels should be preserved in the state it was found. The entire tool should be packaged in protective material then enclosed in plastic. Finally the entire tool should be placed in a wooden or cardboard box and transported to the lab.

Sampling methods may need to be adapted when the scene is outdoors. If the challenge is the recovery of remains, soil samples should be taken at regular intervals up to 100 yards from the gravesite or point of recovery (Saferstein, 2004). Usually a grid search is set up and samples can be taken from each square of the grid and labeled as to which grid it was taken from.

About a tablespoon of soil should be enough for most modern tests. Usually only the surface soil needs to be sampled. The exception to this is in the case of a buried body, In this case soil samples should be taken at regular intervals as the remains are exposed. After the remains are removed from the gravesite the bottom of the grave should also be sampled. It is important to note that a new shovel, spoon or other scoop must be used for each grid and sample. Should the same implement is used serially, uncleaned to recover sample then Locard's Principle is at work again and cross contamination will render your samples useless.

Samples should be placed in plastic vials for transport. If it is not possible to transport the samples immediately they should be allowed to air dry before transport.

A special note for gravesites is that if insect evidence is present the vials should be labeled in pencil rather than pen. The specimens will likely be preserved in alcohol which if it leaks onto the label will destroy ink from pens. Without the label as to the time and place of collection your sample is forensically worthless.

Outdoor scenes often involve vehicles. There are several ways of collecting soil samples from vehicles. Vehicles involved in accidents will sometimes leave lumps of soil from under the wheel wells and fenders on the road way. These lumps should be collected intact and wrapped in protective material to minimize bumping during transport. The purpose behind this is to preserve the layers that have built up to create this lump of soil. A soil analyst can read these layers and know where this vehicle has been. The analyst may also be able to match the layers in a lump of soil to a particular vehicle. I can't help but think this would have been very useful in tracking Ted Bundy or Henry Lee Lucas in their cross county travels.

Clothing and footwear should be collected intact. No attempt should be made to remove soil from clothing, footwear or tires. If these items can be removed intact they should be placed in a paper bag or enclosed in a druggist's fold and then placed in a paper bag. Care must be taken so that the paper bag is protected so that evidence is not lost through holes in the bag. Some evidence can be placed in plastic bags but is must be completely dry. Wet samples placed in plastic with quickly degrade and rot and become useless. Plastic does not "breath" to allow passage of moisture and air. Paper on the other hand will allow moisture to escape preventing rot.

Lumps of soil stained with blood, semen or other biological samples should be collected intact and transported to the lab as dry samples. Any samples containing suspected biological material such as blood, flesh, semen or hair must be clearly labeled so that the analyst at the lab can take precautions to preserve this material. Biological specimens in soil must never be heat dried.

The only time a sample should not be allowed to dry is when insect evidence such as maggots are present. In the case of maggots there is a very specific way of handling this evidence. Samples containing maggots can be placed in aluminum foil with a small piece beef liver and placed in a plastic container. There must be air available in the container.

These samples must be sent to the lab immediately. The specimens can be placed in a thermally protected case such as a cooler. Attempts should be made to protect the samples from extremes of heat and cold. Never use dry ice with biological or entomological (insect) evidence.

In the Laboratory

In the lab items from the victim and suspect should be examined separately. Ideally the items and samples from the victim and suspect should be processed in different rooms. The personnel handling these sample should be assigned to one or the other. If this is not possible then the person handling the material should take extreme care to avoid cross contamination of the samples. The laboratory staff will evaluate the soil samples in several ways. First the mineral content will be tested. Some experienced analysts can moisten a sample and feel the soil. Based on feel alone they can tell the ratio of mineral and organic content. Microscopic examination of soil samples will subsequently reveal the type and nature of the mineral, biological and synthetic content of a sample.

Such talented analysts are not always available. Then, there are several standard testes which can identify the type and origin of the sample.

The first and probably the most common is the density gradient tube. Two different liquids are added to a glass tube in various ratios. Each ratio represents a different density. The soil sample is poured into the tube. When the various particles reach a level in the liquid where their density is equal to the liquid the particles become suspended. This creates a unique profile of bands in the tube which can be matched to other samples.

The samples may also be tested using heat to test the point at which the sample will undergo an exothermic reaction or an endothermic reaction. The sample is heated in a special furnace to various temperatures. In an exothermic reaction the sample essentially burns and releases heat. In an endothermic reaction the sample will absorb the heat. Each sample from different locations will have these reactions at different temperatures according to the mineral, biological and synthetic content.

Electron microscopes can be used to reveal the crystalline structures of minerals and synthetic material in a sample of soil.

Nuclear Resonencing and Mass Spectometry are also methods which may be used in the Laboratory.

It is important that the crime investigators and the laboratory analyst communicate properly. Perhaps this happens best when the laboratory head is knowledgeable about investigations and is the contact person for investigators. The head can be briefed on the questions and issues of the case and then direct the laboratory personnel as to the direction of the inquiry.

The investigators may want to know if sample #12 and sample #76 are similar. The laboratory head will then choose appropriate methods which express the similarity between samples. Once the questions and procedures are chosen everything depends on the integrity of the sample. If they have been collected with care and documented amply then the results from the laboratory can be trusted (Steck, 2004).

Tree Roots Effects on Soil

Trees have a lasting effect on the soil they grow in. Shocking as it may sound, tree roots can also be damaging to the surrounding landscape.

Full of benefits, trees have an enviable position in any landscape, shade, controlling soil erosion, home to many birds, fruits and flowers. Of all the parts of a tree, the roots are perhaps the most unappreciated, as they are unseen.

Roots

There are two types of roots; primary roots that grow deep down vertically into the soil and secondary roots that branch out horizontally. The architecture of the root system is to absorb water and inorganic nutrients and anchor the plant to the ground.

Effects

The roots affect the soil, depending on the type of the tree and the soil. These effects have a direct impact on all the plants grown near the tree. Normally a healthy tree represents healthy soil. A big tree takes up most of the water available in the soil, leaving the other plants dry. Growing as well as mowing lawn grass is another difficulty around a large tree, especially if the roots are protruding outside. Tree roots help control soil erosion, however in some cases the roots have a negative effect on the soil by causing a phenomenon called allelopathy.

Allelopathy

Derived from two words; *allelon* which means of each other and *pathos* which means to suffer. It refers to the chemical inhibition of one species by another, by releasing a chemical affecting the development and growth of surrounding plants. In other words, plants try to get their own space, by restricting other plants from growing too close to them. Allelopathic chemicals secretion are not just restricted to the roots, they are also found in branches, leaves, flowers and fruits. The decomposed leaves and bark affect the top layer of soil, while the roots affect the surrounding soil. The chemical curtails the root growth of other plants by inhibiting their nutrient source, thus influencing their evolution and distribution.

Juglone

It is an aromatic organic allelopathic compound occurring naturally in the roots, bark and leaves of trees in the Juglandaceae family. It releases certain enzymes that inhibits the metabolic function, stunting the growth of many plants and at times even killing an allelopathy intolerant plant. The quantity of Juglone released depends on the weather and soil conditions. The black walnut is the most commonly known for its allelopathic properties. When Juglone sensitive plants come within 0.5 to 0.25 inches of the tree roots, they turn yellow, wilt and die. This in turn, also infects the soil.

Allelopathic Trees

Besides the black walnut, the following are trees with allelopathic properties:
- Sugar Maple
- American Sycamore
- Eucalyptus
- Cottonwood
- Black Cherry
- Red Oak
- English Walnut
- Juniper.

Reducing the Effects of Allelopathy

It would not be sensible to just uproot large allelopathic trees, instead just a few precautions could be taken. A well drained and aerated soil determines the amount of chemical accumulation and the presence of microorganisms in the soil breaks down the toxic chemicals.

Keeping the soil healthy by adding organic matter on regular basis can go a long way in keeping all the plants healthy. Plants can also be grown in containers and pots around the allelopathic tree, just make sure that they get enough sunlight. Another way of ensuring a healthy landscape, is to grow a variety of allelopathic tolerant plants. Some of them are:

- Hawthorn
- Flowering dogwood
- Tulip tree
- Hydrangeas
- Hibiscus
- Daffodils
- Day lilies
- Virginia Creeper.

A tree must be enjoyed, even if some of their qualities are harmful to others. It would be easy to uproot an allelopathic tree or plant, but a good gardener should find ways to grow them all together.

Garden Soil Preparation

We all know that garden soil preparation is one of the vital parts of gardening. This article can provide you with some tips and guidelines about preparing garden soil for planting.

It is a common fact that most plants have some basic requirements to thrive well. Among these requirements good garden soil is one of the primary needs of a plant.

Even though, the soil requirements may vary with individual plants, a good garden soil is considered to be that, which is well drained, deep, loose and fertile. It must have a neutral pH value and must contain decayed organic matter. If you are aware of the factors that can make perfect garden soil, then, you may be able to convert almost all types of soil to ideal garden soil. The following tips and guidelines can help you in garden soil preparation.

What to do Before Preparing Garden Soil for Planting?

First of all, you must know that most of the soil may have some imperfections that can be corrected with some simple procedures. So, before starting with garden soil preparation, you have to check the soil for its imperfections and try some remedy for them.

It has been observed that most people resort to addition of organic matter and gypsum to the soil, so as to correct mild imperfections. Organic matter includes peat, decomposed leaves, pasteurised animal manure, compost, etc. However, it will be always better to test the soil pH levels.

The pH value ranges between 1 (acidic) to 14 (alkaline). While the lower ranges are acidic, the higher ones are alkaline in nature. A pH value of 6 to 7 is considered as neutral and this is what, most plants require.

There are some plants that can tolerate a little bit of variation in the pH range. While, most of the soil imperfections can be rectified by adding gypsum and compost, in some cases, you may need expert advice. If the methods suggested are not cost-effective, you may opt for raised bed gardening or container gardening. Let us now take a look at the various steps of garden soil preparation.

How to Prepare Garden Soil?

Early spring is the best season for starting with garden soil preparation. Once the soil becomes workable, test the soil and gather the materials required for rectifying it. The following tips will help you in this task.

- Start with tilling the soil at the site, where you plan to grow plants. If the area is too grassy, then, make sure to remove the sod too.

Soil Testing : How to Test Soil pH

- It will be better to go for a soil pH test at this stage. The soil has to be dry, while taking samples for testing.
- While, the ideal soil pH for most plants is between 6 and 7, slight variations will be tolerated by most of them. Otherwise, you can correct the soil condition with additives. Also scroll through soil amendments.
- If you are preparing garden soil for vegetables, then the pH should be between 6 and 6.5. Also read through lowering soil pH.
- Once you receive the test results, add those things, that are recommended for rectifying the soil condition. It could be organic matter, gypsum, etc.
- In case of highly acidic soil, you can lower it by adding sulfur, and lime is used to increase the acidity.
- While, peat moss can be added to hard clay soil, rocky soil has to be added with some top soil. Scroll through some information on how to improve clay soil.
- Make sure to remove rocks, weeds and grass, while tilling. You may also break big clumps of soil to smaller parts, but, keep in mind that the soil need not be flour fine.

Now, you are done with garden soil preparation, which holds a vital place in gardening. This is only the basics that can provide an overall idea about preparing garden soil. As mentioned above, some plants need specific soil conditions. So, it will be always better to consult your local nursery or an expert regarding garden soil preparation.

Potting Soil Recipe

The article concentrates on potting soil recipes for a DIY home improvement project. Get step by step instructions on potting soil recipes that will make any garden flourish.

We all know that potting soil is used in gardening and/or different planting systems to give plants a healthy, growing environment. Basically, soil isn't used as an ingredient in making the best potting soil recipe.

There are only additives mixed in the soils which gives the necessary nutrients and fertilizers needed to help grow any plant. These additives also facilitate with water retention and good drainage system. Not only the plants, but the potting soil provides a better grounds for the garden and the container in which the plant is as well. Let's take a look at each recipe for potting soil that are mentioned below.

Potting Soil Recipe #1: Organic Mix

This is an organic soil based mix that yields heavy soil than other peat mixes and has better drainage. To reduce root diseases, compost is known to encourage healthier soil mix. You can use perlite instead of sand and add organic fertilizer on the base. Read more on organic fertilizer recipe.

Potting Soil Ingredients:
- 1/3 screened mature compost
- 1/3 sharp sand
- 1/3 garden topsoil.

Potting Soil Recipe #2: Orchid Mix

Most of the orchids like phalaenopsis and oncidium can be planted with the next potting soil recipe. Although it's a very good idea to repot orchids every one or two years as it improves the health of the plants and helps them bloom.

Soil Testing : How to Test Soil pH

Potting Soil Ingredients:
- 6 parts fir bark
- 1 part peat moss
- 1 part medium grade charcoal.

Potting Soil Recipe #3: Organic Substitute Mix

This next recipe's formula was concocted by the Farm and Garden Project from the University of California, Santa Cruz.

Potting Soil Ingredients:
- 10 lbs bone meal
- 5 lbs blood meal
- 5 lbs ground limestone
- ½ cubic yard sphagnum peat
- ½ cubic yard vermiculite

Potting Soil Recipe #4: Prick-out Mix

This recipe is used when you are planting seeds and then transferring them into the ground when the plant reaches its transplant size.

Potting Soil Ingredients:
- 2 parts sand
- 2 parts aged manure
- 2 parts leaf mold
- 6 parts compost
- 3 parts soil
- 1 part pre-wet and sifted peat moss
- 1 6 inch pot bone meal.

Potting Soil Recipe #5: Tipi Produce Mix

All plant material that comes into any gardens, nursery, or greenhouses can be grown in this mix. This next recipe works well for small and medium plug trays and can grow 1020 flats of peppers, tomatoes, lettuce, squash, cucumbers, and many flowers. You can also fertilize them two times in a week with fish and seaweed fertilizer which is dissoluble.

Potting Soil Ingredients:
- 4 cubic ft. bales sphagnum peat moss

- 1 bag coarse vermiculite
- 1 bag coarse perlite
- 5 parts dolomitic limestone
- 10 parts kelp meal
- 10 parts blood meal
- 15 parts steamed bone meal

Potting Soil Recipe #6: Shade Mix

To grow plants in the shade, follow the recipe to yield an ideal soil. Cover both the humus and soil to wipe out roots, clods and formation of large rocks.

Potting Soil Ingredients:
- 1 part humus (compost)
- 2 parts loamy soil
- 1 part sand

Potting Soil Recipe #7: Cacti Mix

To plant a Christmas, orchid, and any epiphytic cactus plants, take a look at this simple formula that is mentioned below.

Potting Soil Ingredients:
- 3 parts organic potting mix
- 1 part sand
- 1 part perlite

Preparing potting soil recipes at home is always better than buying it from the store and risking if it will work or not. Always select top quality ingredients to make the soil yourself and let your garden prosper. The only difference between a store-bought potting soil and your personal potting soil recipe is your blossoming garden.

Chapter 3
Soil Amendments

To improve the physical properties of soil you add some materials to the soil. This is called soil Amendments. Here's the brief look into types of Soil amendments and its benefits.

Fertility of the soil is important for proper plant growth. Soil amendments are additives to the soil to make the soil fertile so that it provides the necessary nutrients for the crops grown. Amendments are added to improve the physical properties of the soil including water retention, water infiltration, permeability and aeration. The common amendments include fertilizers and soil conditioners. Thus soil amendments enhance the richness of the soil to provide better environment for the roots. Certain amendments are just placed on the layer of the soil. Mulching is an amendment that is left on the soil surface. This is done mainly to prevent evaporation and create attractive appearance. Other amendments are mixed with the soil. Burying the amendments is not a good way to improve the soil performance. These amendments must be thoroughly mixed with the soil to do their work properly. Also you must add amendments in appropriate quantities otherwise it will lead to an adverse effect.

Types of Soil Amendments

There are two categories of amendments - organic and inorganic. The organic amendments are made from natural products. The common organic amendments are sphagnum peat, wood chips, straw, saw dust, compost and manure. The wood chips are mainly used as mulches. These increase the organic constituents of the soil thus providing favourable environment for bacteria and earthworms that enrich the soil. The inorganic amendments are man-made and they include chemicals that are used for making the soil fertile. Though they are used for good output they will reduce the natural nutrients of the soil in the long run.

How to Choose Soil Amendment?

The soil amendments depend greatly on the soil requirements. The types of crops that grow in the soil are important before deciding on the amendment. The other factors include the texture of the soil and its salinity. The amount of amendments depends on the duration you want the amendment to stay in the soil. Before using the amendments, they have to be tested for the organic matter and the pH content. The amendments also depend on your requirements for soil improvement. If you want the soil to improve quickly you have to choose the amendments that decompose at a faster rate. Otherwise composts can be used which decompose slowly. Sandy soils are amended to increase the moisture retaining capacity of the soil. Clay soils are amended to increase permeability and aeration. If you choose the appropriate amendments wisely you can benefit from any type of soil.

Benefits of Different Soil Amendments

You have to add 3 cubic yards of chosen organic amendments per 100 square feet. Some soils many require 4 cubic yards but not more than that. If you add more amendments, it will lead to high salt deposits negating the purpose of using these conditioners. Wood products are added in summer to prevent evaporation and in winter to retain warmth.

These will tie up the nitrogen in the soil. Hence you need to use nitrogen fertilizers to avoid nitrogen deficiency in plants. Sphagnum peat is good for sandy soils. It has high pH and used by gardeners who need acidic soil. Mountain peat is also widely used but they take more time to rejuvenate once they are used up.

Grade 1 biosolids can be used for food crops but they are not used for root crops. As these biosolids come in contact with the roots they should not be used for crops for which the roots are edible. Fresh manure contains high levels of ammonia that may affect the plant. Aged manures or composted manures can be used to solve this problem. By composting to 140F degrees the pathogens can be killed. These composted manures are used for food crops like vegetables.

Land Pollution

Land pollution is basically about the contamination of the land surface and soil of the Earth. Read more about it here.

Land pollution basically is about contaminating the land surface of the Earth through dumping urban waste matter indiscriminately, dumping of industrial waste, mineral exploitation, and misusing the soil by harmful agricultural practices. Land pollution includes visible litter and waste along with the soil itself being polluted. The soil gets polluted by the chemicals in pesticides and herbicides used for agricultural purposes along with waste matter being littered in urban areas such as roads, parks, and streets.

Land Pollution Comprises: Solid Waste and Soil Pollution

Solid Waste: Semisolid or solid matter that are created by human or animal activities, and which are disposed because they are hazardous or useless are known as solid waste.

Most of the solid wastes, like paper, plastic containers, bottles, cans, and even used cars and electronic goods are not biodegradable, which means they do not get broken down through inorganic or organic processes.

Thus, when they accumulate they pose a health threat to people, plus, decaying wastes also attract household pests and result in urban areas becoming unhealthy, dirty, and unsightly places to reside in. Moreover, it also causes damage to terrestrial organisms, while also reducing the uses of the land for other, more useful purposes.

Some of the sources of solid waste that cause land pollution are:

Wastes from Agriculture: This comprises waste matter produced by crop, animal manure, and farm residues.

Wastes from Mining: Piles of coal refuse and heaps of slag.

Wastes from Industries: Industrial waste matter that can cause land pollution can include paints, chemicals, and so on.

Solids from Sewage Treatment: Wastes that are left over after sewage has been treated, biomass sludge, and settled solids.

Ashes: The residual matter that remains after solid fuels are burned.

Garbage: This comprises waste matter from food that are decomposable and other waste matter that are not decomposable such as glass, metal, cloth, plastic, wood, paper, and so on.

Soil Pollution: Soil pollution is chiefly caused by chemicals in pesticides, such as poisons that are used to kill agricultural pests like insects and herbicides that are used to get rid of weeds. Hence, soil pollution results from:

- Unhealthy methods of soil management.
- Harmful practices of irrigation methods.

Land pollution is caused by farms because they allow manure to collect, which leaches into the nearby land areas. Chemicals that are used for purposes like sheep dipping also cause serious land pollution as do diesel oil spillages.

What are the Consequences of Land Pollution?

Land pollution can affect wildlife, plants, and humans in a number of ways, such as:
- Cause problems in the respiratory system
- Cause problems on the skin
- Lead to birth defects
- Cause various kinds of cancers.

The toxic materials that pollute the soil can get into the human body directly by:
- Coming into contact with the skin
- Being washed into water sources like reservoirs and rivers
- Eating fruits and vegetables that have been grown in polluted soil
- Breathing in polluted dust or particles.

How can Land Pollution be Prevented?
- People should be educated and made aware about the harmful effects of littering
- Items used for domestic purposes ought to be reused or recycled
- Personal litter should be disposed properly
- Organic waste matter should be disposed in areas that are far away from residential places
- Inorganic matter such as paper, plastic, glass and metals should be reclaimed and then recycled.

Facts about Land Pollution

What is land pollution? How does it affect the environment? Read all about it in the facts about land pollution.

The process of contamination of the land surface of the Earth is referred to as land pollution. It results from human activities that cause an imbalance in nature. Dumping human and industrial waste, harmful agricultural practices and exposing the land to harmful chemicals leads to the pollution of land. We often ignore the fact that land constitutes soil, which is one of the most important natural resources. While discussing the causes and effects of pollution, we speak of water and air pollution and rarely even think of the adverse effects of land pollution. Let us make an attempt to know some important facts about land pollution.

Facts about Land Pollution

Land pollution is the result of human misuse of soil. Poor agricultural practices, digging up of important resources and dumping of garbage underground can cause land pollution. Urbanisation, the growth of rural lands into urban areas and industrialisation that results in the formation of an industrial society are regarded as the two main causes of land pollution.

The excavation of minerals, the increasing quarrying and mining activities lead to land pollution. The excavation and mining activities lead to the loosening of soil. Increased mechanization leads to the contamination of soil, thus causing severe land pollution.

Deforestation is one of the major causes of loosening of soil, that in turn causes soil erosion. The soil that is left naked on harvesting crops from agricultural lands is vulnerable to being eroded by wind and water. Intensive agricultural practices cause the soil cover to lose its nutritional elements, making it of no use for agriculture.

Excessive use of pesticides and chemical fertilizers causes soil contamination. Chemicals can prove harmful to the animal and plant life. An excessive use of chemicals leads to a decrease in the fertility of soil. Certain herbicides and insecticides lead to toxicity of soil. Fungicides contain copper and mercury, which are extremely harmful to the soil as well as the plant and animal life that thrives in it. Inefficient and unhealthy methods of soil management and harmful irrigation practices lead to soil pollution.

Agricultural and industrial waste, solids from sewage treatment plants, ashes and garbage are other causes of land pollution. The accumulation of inorganic wastes in soil poses a threat to the plant and animal life in that area. Garbage is carelessly dumped into the soil. Non-biodegradable wastes such as plastic and rubber prove lethal to the life in the soil. Plastic and glass bottles, cans, rubber tires and electronic items dumped in the soil make up the main cause of land pollution. Solid wastes are harmful to the terrestrial plants and animals.

How does land pollution affect the environment? One of the major consequences of land pollution is the imbalance in nature, resulting from the harm caused to the wildlife and vegetation on the land. It adversely affects the human respiratory system and results in various skin problems if the toxic materials of the soil come in contact with the skin. The consumption of fruits and vegetables that are grown in contaminated soil can lead to several health hazards in human beings. When contaminated soil is washed away in the water reservoirs, it leads to water pollution, which is lethal to the aquatic flora and fauna. The soil contaminants are driven by the wind, causing air pollution, which is detrimental to health.

It is high time we realise the importance of soil and devise ways to curb land pollution. Maximum use of biodegradable materials and implementation of recycling in order to reuse resources are some of the excellent methods of preventing land pollution. It is important to implement proper methods of disposal of organic waste. It is necessary to educate the masses about the causes and effects of land pollution. We cannot take Mother Earth for granted.

10 Ways to Conserve Soil

Soil is one of the most important natural resources. We need to devise and implement ways of conserving soil. Here is an overview of the 10 ways to conserve soil.

Soil, which is one of the most important natural resources, is often less heeded. The importance of soil conservation is relatively less talked about as compared to the conservation of water and other natural resources. The almost-omnipresent soil is mostly taken for granted. Its omnipresence is ironically the reason behind us, human beings, taking it for a ride. We rarely even think of it as a natural resource that needs to be conserved, a part of the natural wealth that needs to be preserved.

The concept of the conservation of soil takes into account, the strategies for preventing the soil from getting eroded and preventing it from losing its fertility due to an adverse alteration in its chemical composition. Here are some ways to conserve soil.

10 Ways to Conserve Soil

Plant trees: We all know that the roots of trees firmly hold on to the soil. As trees grow tall, they also keep rooting deeper into the soil. As the roots of the trees spread deep into the layers of soil, they contribute to the prevention of soil erosion. Soil that is under a vegetative cover has hardly any chance of getting eroded as the vegetative cover acts as a wind barrier as well.

Terraces: Terracing is one of the very good methods of soil conservation. A terrace is a levelled section of a hilly cultivated area. Owing to its unique structure, it prevents the rapid surface runoff of water. Terracing gives the landmass a stepped appearance thus slowing the easy washing down of the soil. Dry stonewalling is a method used to create terraces in which stone structures are created without using mortar for binding.

No-till farming: When soil is prepared for farming by plowing it, the process is known as tiling. No-till farming is a way of growing crops without disturbing it through tillage. The process of tilling is beneficial in mixing fertilizers in the soil, shaping it into rows and preparing a surface for sowing. But the tilling activity can lead to compaction of soil, loss of organic matter in soil and the death of the organisms in soil. No-till farming is a way to prevent the soil from being affected by these adversities.

Contour plowing: This practice of farming across the slopes takes into account the slope gradient and the elevation of soil across the slope. It is the method of plowing across the contour lines of a slope. This method helps in slowing the water runoff and prevents the soil from being washed away along the slope. Contour plowing also helps in the percolation of water into the soil.

Crop rotation: Some pathogens tend to build up in soil if the same crops are cultivated consecutively. Continuous cultivation of the same crop also leads to an imbalance in the fertility demands of the soil. To prevent these adverse effects from taking place, crop rotation is practiced. It is a method of growing a series of dissimilar crops in an area sequentially. Crop rotation also helps in the improvement of soil structure and fertility.

Soil pH: The contamination of soil by addition of acidic or basic pollutants and acid rains has an adverse effect on the pH of soil. Soil pH is one of the determinants of the availability of nutrients in soil. The uptake of nutrients in plants is also governed to a certain extent, by the soil pH. The maintenance of the most suitable value of pH, is thus, essential for the conservation of soil.

Water the soil: We water plants, we water the crops, but do we water the soil? If the answer is negative, it is high time we adopt the method of watering soil as a measure of conserving soil. Watering the soil along with the plants is a way to prevent soil erosion caused by wind.

Salinity management: The salinity of soil that is caused by the excessive accumulation of salts, has a negative effect on the metabolism of the crops in soil. Salinity of soil is detrimental to the vegetative life in the soil. The death of vegetation is bound to cause soil erosion. Hence, salinity management is one of the indirect ways to conserve soil.

Soil organisms: Organisms like earthworms and others benefiting the soil should be promoted. Earthworms, through aeration of soil, enhance the availability of macronutrients in soil. They also enhance the porosity of soil. The helpful organisms of soil promote its fertility and form an element in the conservation of soil.

Indigenous Crops: Planting of native crops is known to be beneficial for soil conservation. If non-native plants are grown, the fields should be bordered by indigenous crops to prevent soil erosion and achieve soil conservation.

Wind Erosion and Deposition

A short write-up on wind erosion and deposition intended to shed light on how these aeolian processes work to form various landforms on the surface of the Earth. Continue reading for more information on erosive and constructive work of wind.

The erosion of landforms by wind, transportation of eroded material and its deposition are the three attributes of wind activity which facilitate the formation of various landforms on this planet. These three attributes are referred to as the aeolian processes - a term derived from the name of the Greek God of wind *'Aeolus'*. As with various other agents of erosion, even wind plays a crucial role in shaping the surface of our planet by eroding various landforms, transporting this eroded material and depositing it to other places. Some examples of landforms created by wind erosion and deposition are yardangs, deflation hollows, pans, sand dunes, etc.

Aeolian Processes: Wind Erosion, Transportation and Deposition

As with various other types of erosion, even wind erosion - aka aeolian erosion, revolves around detachment, transportation and deposition of soil particles. Wind erosion is most prominent in regions wherein vegetation cover is pretty sparse, and rainfall in minimal. That explains why you get to see very obvious effects of wind erosion in dry, arid regions of the world, wherein wind velocity is very high as there are no trees to obstruct the flow. There exist two methods by which wind carries out its erosional activity on the Earth's surface - deflation and abrasion. Given below are the details of each of these methods.

- Deflation: In geology, deflation refers to erosive action of wind wherein it lifts loose soil particles off the ground and transports them from one place to another. This method of wind erosion is most prominent in deserts, wherein sand particles are lifted by wind and transported to other parts of the desert to form large sand dunes.
- Abrasion: When tiny particles of soil which are suspended in the air are blasted against a standing structure, the standing structure begins to erode over the course of time. This process by which wind erodes various landforms - and results in formation of structures like mushroom rock, is referred to as

abrasion. (When the particles which hit a standing structure break into tiny fragments, it is referred to as attrition.)

These fine particles are then carried by wind to considerable distance depending on what the velocity of wind and the size of particles is. The transportation of these particles is categorised into three different methods suspension, saltation and creeping. Discussed below are the details of each of these methods.

- Suspension: When the diametre of soil particles is 0.1 mm or less, these particles tend to stay suspended in the air and wind carries them along for considerable distance - at times exceeding thousands of miles. This mode of transportation of eroded material by wind is referred to as suspension method.
- Saltation: When the diametre of soil particles is roughly between 0.1 mm and 0.5 mm, they are too heavy to be transported over considerable distance. In such situation, these particles are lifted and deposited at a short distance, and continuous repetition of this method - which is referred to as saltation, transports them to a considerable distance.
- Creeping: When the diametre of soil particles is 0.5 mm or more, it becomes difficult for the wind to lift them and therefore they are transported by wind from one place to another by rolling them along the ground. This method by which wind transports eroded material is referred to as creeping.

The last of the aeolian processes is that of deposition, wherein all the eroded material is deposited by the wind. As with the erosional activity, even the deposition activity of the wind is governed by its velocity. More the speed of the wind, more amount of soil particles can it carry. As its speed slows down eventually - as a result of change in the pattern of landforms pr presence of vegetation, wind starts depositing the eroded material on the ground. Heavy particles are dropped first, while light particles are transported further and deposited when wind speed decreases by a significant extent. While yardangs, blowouts, ventifacts, mushroom rocks, hoodoos, etc., happen to be landforms created by wind erosion, the deposition of this eroded material results in formation of sand sheets, sand dunes, sand hills, loess, paha, ergs, etc.

While water as an agent of erosion is way more powerful as compared to wind when it comes to regions which experience a significant amount of rainfall, it is wind that is the lone powerful force when it comes to arid landscapes devoid of water. However, that doesn't mean wind erosion and deposition only takes place in deserts and other arid regions.

Even coastal areas - with wind blowing over the ocean at a considerable speed, are vulnerable to aeolian processes with several beaches and sea cliffs of the world showing the traces of the same.

Soil Erosion Solutions

Soil erosion is a growing concern with every passing day. We can't just sit back and do nothing about it. It needs to be curbed, and for that we need soil erosion solutions. Let's hope what follows is of some help...

All through our growing up years we studied it in school, perhaps even heard our folks make us aware of this geographical concern that we're faced with.

Soil erosion has most certainly been of relevance in terms of environmental problems that we're facing today. So what would one do in a case like such? Its simple, soil erosion solutions are the need of the hour. But before getting to that, what is soil erosion? Very briefly, let's go over that first.

Soil Amendments

What is Soil Erosion?

To begin with, it is important to understand that soil erosion is a natural process which has been taking place since time immemorial, but has increased considerably in the recent years due to absolute misuse of land.

A type of soil degradation, soil erosion is a process where soil is naturally removed by the action of either wind or water. Extreme or frequent rainfall, storms of great intensity, lack of vegetative cover, etc., are some of the common reasons leading to soil erosion. But why is it so important to suppress this natural process? Its simple... Soil erosion has several negative effects. Some of them are, loss of productive farmland and agricultural yield, road damage, and contamination of water. Needless to say then, soil erosion prevention is extremely essential. But how does one go about it? We've all heard 'a journey of a thousand miles begins with a single step', so that is exactly the best way to get your foot into soil erosion solutions.

Soil Erosion Causes and Solutions

We went over what soil erosion is, but it was hardly detailed enough. Let's get a better look at its causes, before we go over the solutions.

Causes

There are two possible causes which contribute towards erosion. Soil erosion occurs when there is a major impact of natural causes, and the other reason is those brought on by humans. A list of some such causes will help for better comprehension.

- Climate: Climatic conditions are always a large cause of environmental problems faced by us. Reasons such as temperature, relative humidity, frequency / intensity / distribution of rain or storms, and precipitation are a few such conditions.
- Flow of Water: Water is another common cause of soil erosion. This factor works in two ways, flow of water on the surface, and underground flow of water. The direction of the flow, along with its velocity also matter when it comes to soil erosion due to water.
- Soil Type: As much as a person may want to dismiss this option, the type of soil is a large contributing factor to soil

erosion. Certain kinds of soil are less resistant to erosion due to their physical and occasional chemical properties.
- Human Induced: This reason should come as no surprise. Reasons such as deforestation, construction, incorrect methods of farming, grazing, etc., are just a few in the list of soil erosion due to man.

Solutions

Studies say that only 11% of all the land available can be used as agricultural or cultivable land. In a case like this, it seems only logical that once the causes have been determined, solutions must be figured out too, so it's time to move on to them now...

- Restricted Construction: Construction seems like a massive problem these days. Any spare bit of land we find, we feel the need to construct something. Little do we realise that immense urbanisation will soon leave us with nothing to care about.
- Grazing and Vegetative Cover: Taking care of these two aspects is yet another solution to soil erosion. Dense vegetation is a great method against erosion as the roots of plants help hold soil and soil particles together. This cover also prevents the wind from acting against it.
- Use of Mulch: This is a good method of prevention, provided it is used on flat land, and not slopes. Since mulch is made of things such as straw, shredded bark, and other natural materials, they allow gradual soaking in of soil, hence, preventing erosion.
- Use of Retaining Walls: This method may not necessarily qualify as extremely effective in terms of major prevention, but it most certainly can control erosion to a certain extent. Retaining walls made of brick or mortar can help around areas like vegetable patches and the likes.

Other useful methods are, the use of wind breakers, gardening, healthy soil, etc. Hopefully these soil erosion solutions will help you the next time you decide upon it. We need to take care of our planet, after all it's the only one we own.

Soil Erosion Facts

Soil erosion facts are associated with the nature and causes of this natural phenomenon. It's a major environmental concern that everyone should be aware of. Read ahead to know the details.

Soil Amendments

Like other environmental concerns, soil erosion is also one of the most damaging factors of the environment. Soil erosion facts are associated with the nature of erosion, the causes and the preventive measures. We shall begin with the basic question, what is soil erosion. In brief, soil erosion is the process of detachment or weathering away of the soil particles from ground by various agents, rendering the land barren. You shall come to know the details of this geological process from the following segment.

Interesting Facts about Soil Erosion

Soil erosion has become a topic of concern by researchers and environmentalists. Sincere steps of soil erosion prevention are taken by government to curb its disastrous effects. An intense erosion will deprive the land of any vegetation. This further results in minimal rainfall and drought. Thus, the natural cycle is disrupted. Get the detailed information about soil erosion from the next segment.

Causes of Soil Erosion

- Soil erosion facts are associated with the various agents of nature causing erosion naturally.

- The high speed of the flowing river water is one of the prime causes of soil erosion of the river basins and the coastal regions. Areas lying on high altitudes are eroded and the sediment gets deposited on the low lying lands.
- Wind erosion is the most devastating agent of soil erosion. It's more evident in the dry areas like desserts. The power of the wind has the ability to weather rocks, soil, etc and transport them to a different zones, leaving behind a dry land.
- Soil erosion can also be caused by the glaciers and ice. This type of erosion is more evident in snow capped regions and high altitudes, where soil particles get removed by the movement of glaciers through them.
- Soil erosion causes are also associated with climatic conditions like temperature, wind speed and rate of precipitation affecting the intensity of erosion. A land grazed by farm animals is more prone to soil erosion.

Effects of Soil Erosion

- One of the key facts about soil erosion is creation of a new type of land where fresh deposition of soil particles will take place.
- The vegetation of the region is affected as a result of soil erosion. When the soil gets washed away then the productivity or the fertility of the land decreases.
- The moisture and mineral holding capacity of the soil is highly reduced. Thus the land becomes devoid of any type of agricultural activities.
- This gradual process reduces the weight of the earth's mantle and the surface layers, which might subsequently lead to tectonic shift in the earth's crust.
- Barren lands results in occurrence of drought due to absence of rainfall in those particular regions. These were some of the causes and effects of soil erosion.

Control of Soil Erosion

- Soil erosion prevention measures are implemented on a large-scale today. To name a few, they are, shifting vegetation, afforestation, conservation tillage, gardening and contour farming.

- Erosion control products help in retaining the organic value of the soil as well as the pH. The requirements of the soil are restored by mulching.
- Soil erosion control is also carried out by planting erosion control plants. Native plants, willow trees, yarrow flowers are help to retain the moisture of the soil.
- One of the most effective technique to prevent soil erosion is by controlling the rate of grazing in a land vulnerable to erosion.
- Regular watering of soil also keeps loose particle attached with the ground. Thus, preventing it from getting eroded.

I hope you have derived valuable information from the article 'soil erosion facts'. It's a serious environmental concern that everyone should be aware of. Once the facts are clear, we can come forward to effectively implement the steps to control soil erosion.

What is Soil Erosion

Erosion of the soil is very common in areas where there are steep slopes and less trees. One of the best examples of soil erosion is the Himalayan mountains in Nepal. Due to the sudden rise in human population in the hilly region, soil erosion has occurred causing many areas to suffer from drought.

Soil erosion is never beneficial for the land and has high negative impacts. Soil contains many nutrients and minerals which are helpful for crops to grow. So let's understand in detail what is soil erosion and its various causes.

What is Soil Erosion Caused By?

There are many factors which cause soil erosion and hence makes the productivity of land suffer. So let's have a look the main causes of soil erosion.

Water Erosion

Rain can be quite a pain for the farmers when they plow the soil. Rains coming at 40 mph drains out seeds and even washes out the soil. If the cultivating fields is on the top of a hill, then the soil is splashed down which causes more problems. Water erosion can be divided in three categories; sheet erosion, rill erosion and gullies. In sheet erosion, rainwater removes thin layers of soil from the land surface. Rill is the most common form of soil erosion. It occurs when soil is taken by water is small streams and which causes poor surface draining. You can see gullies of muddy water because the streams have accumulated and formed a small puddle.

Wind Erosion

Wind erosion is one of the main soil erosion causes. It occurs when bare fertile soil is exposed due high velocity blowing winds. When the force of winds is higher than the gravitational force of the soil particles, wind erosion takes place. According to various professors of environmental studies, wind moves light soil particles by bouncing or hopping and heavy particles by rolling down the surface. Most of you don't know but wind erosion is one of the most common and visible forms of soil erosion. Dust storms are the perfect example of wind erosion, they also cause potential damage to public utilities and are also responsible for the growing number of accidents on the American highways.

Human Erosion

The third and the most important cause of soil erosion is us; humans. Since decades we have indulged in hazardous activities like deforestation, overgrazing and improper plowing which have led to ecological imbalance and soil pollution. Did you know that roots of several trees, plants, shrubs and grasses help in soil erosion prevention. Cutting of trees especially on hill slopes causes soil erosion to a greater extent. Human population is expanding and we are taking up all the land which were once home to millions of trees and plants. I think it's safe to say more than the natural causes, humans are responsible for soil erosion.

Soil Amendments

Facts about Soil Erosion

Now that you've understood what is soil erosion and its various causes, let's learn some more soil erosion facts which will help you understand its drastic effects of soil erosion on our planet.

- Soil erosion basically occurs when soil is washed away by rain or blown away by wind. Wind and rain cause a significant amount of soil erosion every year.
- The removal of soil from one area to another can cause the weight of the lower crust and mantle of the earth to lighten down. In future this can cause a tectonic uplift in areas of high soil erosion.
- One of the most astonishing facts about wind erosion is that it causes displacement of porous and permeable soil which gives rise to calamities like drought and poor vegetation.
- Soil erosion often fills up lakes and streams which are used for water supply thus increasing the cost of water treatment.
- There are many soil erosion control methods which can be implemented to prevent soil erosion to a large extent. The best way to stop soil erosion is to increase the growth of various plants and trees on steep slopes as these areas are more susceptible to erosion.

Remember, soil erosion can be reduced by building terraces on hill slopes and implementing irrigation schemes which are specially designed to eliminate drought. However, the best and the most effective way of preventing soil erosion is afforestation.

Soil Erosion Control

Much has been said, written, and debated about one of the biggest concerns of our environment - soil erosion. However, not much has been done to control the same. What affects the surface of the earth as a result of loss of soil due to wind and water currents is accelerating with high speed today, leaving no facet untouched.

While deforestation, industrialisation, and other destructive activities of man are some prime soil erosion causes, if not controlled, it can deplete the soil cover of our planet to an extent we can't even imagine. All things considered, this article stresses on what are the various soil erosion control methods that can be incorporated to save our planet from this calamity.

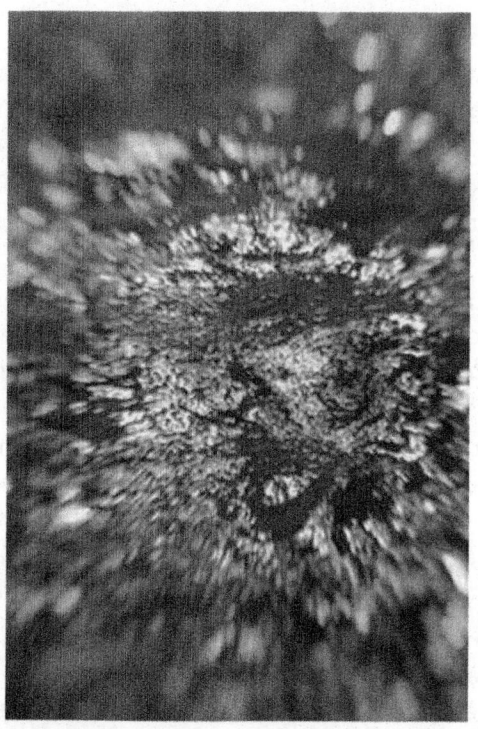

Soil Erosion Control Methods

Soil management is possible only when soil erosion prevention practices are taken sensibly and seriously. Gardening, contour farming, use of retaining walls and fertilizers can contribute to effective soil erosion control on a regular basis. However, for larger areas that are dedicated to agriculture and farming, land investment is one thing that can improve the structure of soil in a number of ways. It will not only sustain crop yields, but will also keep up the nutrient content of the crops. Moreover, with reduced hazards while carrying out the soil erosion and sediment control, water quality will improve too. Methods may be many, but prevention is the utmost concern. Following are some soil erosion control approaches that can be acquired to restrain and prevent soil erosion to a great extent. Have a look:

Soil Erosion Control Products

Quality soil erosion control products have been introduced by global manufacturers to contribute to the increasing soil erosion. These products for landscaping are highly effective, and are extremely environment-friendly. For example, some manufacturers have created specific connectors that are inserted into the soil, thereby, increasing the strength of the soil structure as well as soil pH levels. Not only do they increase vegetation by enhancing the soil structure, they also satisfy all the requirements of soils of even the most difficult mountain slopes. These products are a breakthrough in the field of soil erosion control, and have been operational in the form of soil erosion control mats and fibres. Soil stabilisation is no more a difficult task, and these products are easy to use.

Soil Erosion Control Plants

Specific plants are considered to be a great way of controlling soil erosion in various areas. One of the most popular erosion control methods is streamside planting as a result of which, the plants that act like a net on the soil, hold it all together. Rainfall is much likely to occur in places where vegetation is rich, and hence, while rain would gradually drool down the forest, soil erosion would be reduced to a large extent. A number of flowering plants dedicated to soil erosion control are much acquired for use. For example, *native plants* are the most preferred choice for enduring erosion prevention and control. *Willow trees* which act as effective soil erosion control mats and fences, along with *Blue Rug Juniper* plants are other good options. Furthermore, *Yarrow flowers* and native grasses provide excellent slope protection, and contribute to reduction in soil erosion.

Other Control Measures

Apart from the above methods of soil erosion control, there are methods like *mulching, intercropping, agroforestry,* and usage of *crop residues* that can prevent the soil from getting eroded, and improve the structure of soil, in case of soil erosion. Mulching is covering the soil with a layer of organic matter that helps in keeping the soil moist, and enriches the quality too. On the other hand, intercropping and agroforestry are some measures that render protection to the soil from rain splash. Lastly, using crop residues onto the soil post harvesting protects the soil from getting eroded until the next crop development.

Even though there are certain measures available for soil erosion control, prevention should always be an upper priority. In order to reclaim a pathetically eroded land, only those methods should be incorporated that contribute to increasing soil fertility. It is highly recommended not to use all the soil conservation methods on a single piece of land. Instead, use one of the breakthrough methods to get maximum results. Proper execution and preventive measures can help in sustaining the much degrading quality of our environment.

Wind Erosion

It is known that wind erosion is a natural geological factor, without which the earth wouldn't be called earth, because this planet *too needs some changes*. Infringe into this article, the definitions, causes, effects and other parameters that comprises this erosion.

The purpose of this article is to observe the effects of wind on sand as in desserts and on sandy coastlines. Wind erosion is defined as the transportation and re-deposition of soil particles by wind which in turn segregates the fine sand and smaller particles from an arable land. Wind erosion usually takes place on a land which is bare, loose, cohesive, dry or finely granulated, thus causing a removal of the finest particles of soil and devising the fertility of the soil to degrade. This action is also known as aeolian (or eolian) which means erosion (it was named after the *Greek God 'Aeolus'*).

When the wind is blown in the sand and manages to hit an obstacle, the sand tends to pile up into a ridge or a mound and is called a *dune*. Such dunes by the wind can be actuated grain by grain. Well, it's seen that wind erosion has become major problems in most parts of the world, particularly in the arid and semiarid regions it has worsened even more. Agricultural practices that comport tilling and sowing are disturbed by the erosion of the soil in large quantities. Places like Australia, North and South America, Africa, eastern, central and southern parts of Asia, etc., are the major victims to lose the balance of the ecosystem. Let's dig deeper into this article and see besides the action of picking and moving of objects and materials by the wind, what other facts and factors are involved under this aspect of environmental science.

The Causes and Effects

Generally, the process of wind erosion is a natural phenomena, but it can sometimes be accelerated by the influence of humans also, especially by factors of overgrazing, agriculture, cultivation and urbanisation. Inducing deforestation to construct a civilization or undiscriminating the land for an agricultural or cultivation purpose, etc., exposes the soil to an increment of wind erosion. Along with this gets affected are the microorganisms and living things that live deep inside the soil and are known to be the benefactors for the human race. They help in the soil aeration and structure of the land.

The condition of wind erosion (keeping the particle size into consideration) is that, it has three various modes of particle transportation, along the surface - creep, saltation and short-term and long-term suspension - i.e. hours and days. Also, writing on the wind erosion facts, the main factor is the velocity of the wind. Basically, there are two favourable conditions to the occurrence of wind erosion: one - the meteorological conditions (i.e high wind speed) and two - the ground conditions. If the land or soil is susceptible with no residue of crops or plants on it, there is an erosion of the texture and organic matter from the surface of the land. Such a process of soil erosion leads to a massive imbalance between the structure of the soil and the roots of the crops. Well, there are plenty other influencing parameters to the causticity of the soil too, like roughness, vegetation cover, filed size, duration and frequency of wind (velocity, humidity, direction) radiation, rock type, evaporation and also precipitation. Hence, now we know, there is not one reason but several different elements involved which affect the geological posit.

After knowing the wind erosion facts and all its causes and effects, we shall see the names of different places where wind erosion has changed the visual aspect of the ecosystem on earth.

Wind Erosion Examples

- *Antarctica* - Most of the boulders in this place has been eroded by the prevailing winds.
- *Wisconsin* - There is a large basin which is deflated near the Spring Green by the flood plain of the Wisconsin River.

- *Michigan* - There are coastal dunes a.k.a beach dunes on the shore of Lake Michigan.
- *Colourado* - The dunes in the Great Sand Dunes National Monument, located northeast of Alamosa, are the highest dunes in U.S. which are over 200 metres high.
- *Nevada* - The dune east of Fallon in Nevada are formed by sand and wind blown miles across the valley. It is known that funneling of the wind takes place through the low pass but is unable to carry the sand with it.

Under the roof of this world there are too many aspects which are unheeded by the human race. Simple changes in geological conditions can affect us or sometimes we become one of the causes of their occurrence. From this article, it was interesting to see the increase in the occurrence of wind erosion due to several factors that play an unnoticed important role everyday, that too in the same environment we breathe in.

Soil Erosion Prevention

Soil erosion prevention is very important for our sustenance, and we must understand the role we play in causing soil erosion, so that we can learn how to prevent it.

Soil Amendments

Soil erosion, though a natural process, has had a devastating impact on the earth. In a majority of cases, the process has been speeded up by human activity. The disturbing facts about soil erosion are these; it impacts the growth of crops, vegetation and plants as the soil no longer contains enough nutrients to assist a healthy growth. Having examined the gravity of the situation, it is evident that a collective effort is needed to be made towards soil erosion prevention.

Soil erosion is basically the shifting of soil particles due to rain, wind, floods and melting ice. We as individuals can do very little to extend soil erosion control when it is caused by nature. A significant amount of soil erosion takes place by rivers during floods or because of storms, droughts and melting ice. However, this kind of erosion can only be tackled at the national level. Humans contribute to the process of erosion by cutting down forests, undergrowth and excessive watering of gardens or surrounding areas. Since we have no control over the wind or rain, we can concentrate on one simple aspect which is to increase plant life around us which holds top soil in its place.

Soil Erosion Solutions

Here are some simple measures we can take to reduce soil erosion:

- Gardening: Take up gardening as a hobby and get in touch with nature. The grass and trees you plant or even the fences you put up will prevent heavy rain from beating down on your land and keep the top soil in place. It will also prevent the soil from drying up and getting blown or washed away.
- Conservation tillage: Often used by farmers but can be adapted by gardeners as well, it is simply leaving some vegetation on the ground instead of stripping the entire area bare. In its simplest form it involves planting a short hedge around each plot of flowers or vegetable patches or even leaving a strip of grass between every few beds of flowers/vegetables. This will help to keep the soil in check.
- Keep Soil Healthy: This is an important step to prevent soil erosion. Use fertilizers, manure or compost regularly, soil thus treated becomes rich in organic matter, binds together and is less prone to being washed or blow away.
- Plant Wind Breakers: In areas prone to gusty winds, plant wind breakers in the form of trees, hedges and bushes or even put up a wooden/plastic fence.

- Contour Farming: If you happen to be living in the hills, this method is very useful in preventing soil erosion by slowing down the flow of water down the slopes. It is done by following the natural contours of the land while planting.
- Avoid Leaving Land Barren: On land that is not in use one should plant a 'cover crop'. This helps your land in two ways; to begin with it prevents soil erosion and also provides nutrients for the land in the form of nitrogen fixers if certain cover crops are used.
- Matting: One of the commonly used products in soil conservation is 'matting'. Readymade matting made of wood fibre is commonly used in household vegetable gardens and vacant plots. The matting which is placed on top of the soil prevents soil erosion while at the same time it allows plants and trees to grow through it.
- Use of Retaining Walls: Very useful in areas where rains is heavy and water erosion affects the soil, one can prevent it by constructing small retaining walls around your vegetable plots. You can use bricks and mortar or even pack in heaped soil in rows. Know more about the types of retaining walls.
- Use of Mulch/Fertilizer: Another useful method is applying a layer of mulch and fertilizer over the soil. This prevents the rain from beating down hard directly on the soil at the same time water slowly soaks through the soil and enriches it. The mulch and fertilizer layer helps the soil to regain its PH levels.
- Avoid Excess Watering: While it is important to keep the soil moist which at times may be difficult, like in summers, avoid over-watering as this washes away the top soil and degrades the land.

Soil Conservation Methods

Soil defines the ecology around it, and hence it important for us to opt the soil conservation methods. Let us look at the soil conservation practices, used around the world.

Soil conservation is maintaining good soil health, by various practices. The aim of soil conservation methods is to prevent soil erosion, prevent soil's overuse and prevent soil contamination from chemicals. There are various measures that are used to maintain soil health, and prevent the above harms to soil. Here are the soil conservation methods which are practiced for soil management.

Soil Conservation Strategies

There are many ways to conserve soil, some are suited to those areas where farming is done, and some are according to soil needs. Here are the various soil conservation practices.

Planting Vegetation

This is one of the most effective and cost saving soil conservation strategy. This measure is among soil conservation technique used by farmers. By planting trees, grass, plants, soil erosion can be greatly prevented. Plants help to stabilise the properties of soil and trees act as a wind barrier and prevents soil from being blown away.

This is also among strategies used for soil conservation methods in urban areas, one can plant trees and plants in the landscape areas of the residential places. The best choices for vegetation are herbs, small trees, plants with wild flowers, and creepers which provide a ground cover.

Contour Plowing

Contour farming or plowing is used by farmers, wherein they plow across a slope and follow the elevation contour lines. This method prevents water run-off, and thus prevents soil erosion by allowing water to slowly penetrate the soil.

Maintaining the Soil pH: The measurement of soil's acidity or alkalinity is done by measuring the soil pH levels. Soil gets polluted due to the addition of basic or acidic pollutants which can be countered by maintaining the desirable pH of soil.

Soil Organisms

Without the activities performed by soil organisms, the organic material required by plants will litter and won't be available for plant growth.

Using beneficial soil organisms like earthworms, helps in aeration of soil and makes the macronutrient available for the plants. Thus, the soil becomes more fertile and porous.

Crop Rotation Practice

Crop rotation is the soil conservation method where a series of different crops are planted one after the other in the same soil area. This method is used greatly in organic farming. It is done to prevent the accumulation of pathogens, which occur if the same plants are grown in the soil, and also depletion of nutrients.

Watering the Soil

We water plants and trees, but it is equally important to water soil to maintain its health. Soil erosion occurs if the soil is blown away by wind. By watering and settling the soil, one can prevent soil erosion from the blowing away of soil by wind. One of the effective soil conservation ways in India is the drip irrigation system which provides water to the soil without the water running-off.

Salinity Management

Excessive collection of salts in the soil has harmful effects on the metabolism of plants. Salinity can lead to death of the vegetation and thus cause soil erosion, which is why salinity management is important.

Terracing

Terracing is among one of the best soil conservation method, where cultivation is done on a terrace levelled section of land. In terracing, farming is done on a unique step like structure and the possibility of water running off is slowed down.

Bordering from Indigenous Crops

It is preferable to native plants, but when native plants are not planted then bordering the crops with indigenous crops is necessary. This helps to prevent soil erosion, and this measure is greatly opted in poor rural areas.

Soil Amendments 113

No-tilling Farming Method

The process of soil being plowed for farming is called tilling, wherein the fertilizers is mixed and the rows for plantation are created. However, this method leads to death of beneficial soil organisms, loss of organic matter and compaction of soil. Due to these side effects, the no-tilling strategy is used to conserve soil health.

These were the 10 ways to conserve soil used across the world. Soil is a very important constituent, and is developed by a long process of weathering and disintegration of rocks which turn into sand or clay. The clay like fertile soil provides home to organisms like earthworms, beetles, ants which live in it. Soil provides anchorage to plants and trees. The plants and trees provide home to birds and animals. The crops growing on the soil provide us food and clothes. Thus, soil defines the quality of life around it, which is why it is important to use these soil conservation methods. Branches of environmental science like Earth science are constantly trying to find new methods, for maintaining the ecological balance. In different parts of world people studying soil science, are coming up with different new beneficial soil conservation techniques.

Chapter 4

Soil Conservation

Soil conservation is a very important issue, both in developing nations where a good portion of income is derived from agriculture and in developed nations where mechanized farming and an overabundant use of chemical fertilizers can sometimes have a detrimental effect on the land. But soil conservation is not only for farmers and agriculturists - it has a far-reaching effect on the environment and so concerns all of us. There is a growing need to prevent and control soil erosion and soil contamination, and to maintain soil fertility. Implementing various soil conservation strategies and methods can help in stemming erosion of the soil, in preserving the quality of the soil and in increasing its productive capacity. Good soil conversation leads to enriched lands, better crop yields, good financial returns and a balanced environment.

In order to carry out effective soil conservation, you must first consider several aspects of the land in question. Here are some of the things that must be considered -

Soil Conservation

- Is there enough vegetative cover over the land or does it need to be developed?
- What are the proper erosion control methods that can be utilised and how will you implement them?
- What is the soil salinity level and how are you going to manage it?
- What is the soil acidity level and how are you going to control it?
- What is the soil mineral content and do you need to regularly add mineral supplements?
- Is the soil contaminated and, if so, what methods are you going to use to remediate it and to prevent future contamination?
- Are there beneficial soil organisms in the soil and are they allowed to thrive?

Once you know what kind of soil conservation is needed, you can consider implementing any or all the following methods -

- Planting dense rows of trees as wind-breaks along the borders of the land, especially on the sides that are exposed to stiff winds. Their roots stabilise the soil and prevent it from being shifted away by the wind.
- Planting crops in rotation. Alternating crops helps avoid depleting the soil nutrients too fast.
- Planting cover crops. These stabilise the soil and reduce the effects of soil erosion. They also discourage the spread of weeds and help the soil retain its moisture in the summers.
- Plowing along the contour of the land.
- Planting crops parallel to the slope of the land.
- Going for strip cropping. This involves planting grasses or pulses between regular crops like corn. The corn crop is not particularly effective in preventing soil erosion, but the grasses keep erosion in check.
- Adding mulch to the soil surface. This prevents erosion by acting as a barrier and catching run-off water.
- Adding coir logs as barriers. These are very useful in areas where too much erosion has taken place and act as erosion preventing barriers and a support for new developing

vegetation. Aside from coir barriers, sand bag and gravel bag barriers are also used.
- Growing grass on slopes and in waterways. Grassed waterways prevent too much soil from being washed away.
- Making use of natural as well as man-made fertilizers.
- Keeping the land fallow in order to rest it.
- Managing the levels of salinity. In areas where irrigation is in excess or where the saline water tables are low, the salinity levels can go up and make the land unsuitable for agriculture. This problem can be resolved by the use of humic acid.
- Managing the soil pH. The soil pH is what determines the amount of nutrients that the plants can absorb from the soil. Soil pH levels can be raised or decreased, as needed, by adding certain chemicals - for example, agriculture lime for raising pH level and ammonium phosphate for reducing it.
- Encouraging beneficial soil organisms like earthworms and nitrogen fixing bacteria to thrive in the soil. The presence of such organisms enriches the soil.
- Using man-made chemical insecticides, pesticides and herbicides in very low amounts. Over use of chemicals can poison the soil and kill of useful organisms, and is generally harmful to the environment.
- Regularly add minerals to the soil. Minerals provide much-need nutrients to the soil to be absorbed by the plants. To mineralise the soil, add chemical supplements or try adding crushed rock.

Ways to Conserve Soil

Trees and forests are being felled at an alarming rate. This has led to large-scale erosion and depletion of soil. As a result of this, soil conservation has become a critical issue needing immediate attention. This article looks into the ways to conserve soil.

Soil sustains all life on earth. Unfortunately this vital resource is being depleted rapidly. Mankind must take instant and strong steps to conserve soil.

Soil is made up of organic and mineral matter that sustains plant life and shelters organisms responsible for nutrient cycling

Soil Conservation

What are the different Types of Soil?

There are different types of soil. They are as follows:

Clay - This soil is smooth and silky.

Sandy soils - It contains tiny amounts of quartz and silica and less than 10% clay.

Loamy - This soil is a mixture of sand and clay with varying proportions.

Calcareous or chalky soils - This soil may contain limestone or chalk and very little plant food.

Peat soil - This soil is usually found in marshy land and is a source of fuel. It contains more than 20% humus.

Plants and trees hold the soil together. When plants and trees are cut down the soil becomes loose and is carried away by wind and water. When it rains heavily the rainwater cannot penetrate the soil and the rainwater carries the soil particles away. This is known as soil erosion. Common types of erosion are gully erosion, rill erosion, sheet erosion and rain-splash erosion. The Grand Canyon in America was formed by soil erosion. There is more erosion on steep slopes and it also depends on the type of the soil

Some ways to conserve soil are as follows:
- Use rotational grazing. There is a short grazing period followed by a rest period of longer duration. Grazing is done when the farm is still in the vegetative stage. This prevents crops from being totally eaten away.

- Change the plant species on your farm. Don't use tilling, instead use herbicides to kill the existing plants.
- Cover the entire soil with plants, which will significantly reduce erosion.
- Along with the plants it's important to water soil, which keeps it damp and makes it settle down.
- Don't cultivate soils on steep slopes. Do terrace farming.
- You can construct wind barriers at the boundaries of the farm. This will prevent wind from blowing away the soil.
- Add humus to your soil. It will prevent soil erosion.
- Keep grassed waterways to drain out storm water.
- To fight storm water, use structures made of natural materials. Use logs or collection of large stones instead of cement and concrete. Natural resources are more effective and inexpensive.
- Plant strips of grass, trees or shrubs between water and cropland. It prevents surface movement of fertilizers, pesticides and soil. Strips cause increased runoff of water into soil. There is increased denitrification, in which microbes convert nitrate-nitrogen into gas form that dissipates to the atmosphere. It decreases the amount of nitrate available to move into groundwater and surface water supplies. They absorb nutrients, sediment, and pesticides moving from adjoining cropland before they reach the water sources. The strip's trees, grass or shrubs absorb the nutrients and pesticides.
- Farm taking into account the shape of the land. The small grooves and channels that you fashion play the role of dams, trapping runoff water, sediment, nutrients, and pesticides, and pushing them along graded crop rows to outlets such as grassed waterways.

So we see that there are a number of simple and effective ways by which we can conserve soil. Only a global and earnest effort can save this life sustaining resource. So do take the necessary steps and fulfill your obligation towards your planet.

Soil Erosion Causes

The geological process wherein soil particles are detached and transported by the various agents of erosion is known as soil erosion. These agents of erosion include water, wind, glaciers, and gravity.

Soil Conservation

There are plenty of soil erosion causes which are triggered by various phenomena taking place on the planet. They may range from slow moving ice bodies in the glacial mountains, to landslides caused by earthquakes. Before we move on to the main causes of soil erosion, let's take a brief look at the various types of soil erosion.

Types of Soil Erosion

Soil erosion is broadly categorised into different types depending on the agent which triggers the erosion activity. Mentioned below are the four main types of soil erosion.

Water Erosion

Water erosion is seen in many parts of the world. In fact, running water is the most common agent of soil erosion. This includes rivers which erode the river basin, rainwater which erodes various landforms, and the sea waves which erode the coastal areas. Water erodes and transports soil particles from higher altitude and deposits them in low lying areas.

Wind Erosion

Wind erosion is most often witnessed in dry areas wherein strong winds brush against various landforms, cutting through them and loosening the soil particles, which are eroded and transported towards the direction in which the wind flows. The best example of structures formed by wind erosion are mushroom rocks, typically found in deserts.

Glacial Erosion

Glacial erosion, also referred to as ice erosion, is common in cold regions at high altitudes. When soil comes in contact with large moving glaciers, it sticks to the base of these glaciers. This is eventually transported with the glaciers, and as they start melting it is deposited in the course of the moving chunks of ice.

Gravitational Erosion

Although gravitational erosion is not as common a phenomenon as water erosion, it can cause huge damage to natural, as well as man-made structures. It is basically the mass movement of soil due to gravitational force. The best examples of this are landslides and slumps. While landslides and slumps happen within seconds, phenomena such as soil creep happen over a longer period of time.

What Causes Soil Erosion?

There is no particular soil erosion cause which can be singled out and assumed as the main cause of soil erosion. The process has many underlying factors, some induced by nature and some by humans.

Human Induced Causes of Soil Erosion

Human exploitation of nature is perhaps the most hazardous cause of soil erosion, which has increased over the last decade. Human activities, such as faulty farming systems, deforestation caused by overgrazing, clearance of land for agricultural purposes and construction, dam construction and diversion of the natural course of river, and mining activities are just a few among the various human activities which have either directly or indirectly weakened the topmost layer of the planet, thus making it vulnerable to excessive wearing away by the various agents of erosion. For instance, tree roots help in holding the soil together, and therefore depletion of vegetation cover is bound to make soil vulnerable to erosion by running water.

Natural Causes of Soil Erosion

Besides the above human causes of soil erosion, there are a few natural causes of erosion as well. The major natural factors that can cause soil erosion include:

Gradient of Slope: Gradient of the slope is an important factor when it comes to soil erosion. In fact, erosion and gradient have a direct relationship.

The steeper the gradient, higher is the rate of erosion and vice versa. This factor plays an important role in water erosion, glacial erosion, and gravitational erosion.

Soil Properties: The vulnerability of a piece of land to soil erosion depends on the physical and chemical properties of the soil as well. Different types of soil have different physical and chemical properties. The texture, structure, water retention capability, etc. play an important role in determining whether the soil is susceptible to erosion by various agents of erosion or not. This factor is common in all the above mentioned types of erosion.

Water Flow: Hydrological cycle, especially the surface flow as well as underground flow also play a major role in soil erosion. Variation in the velocity and type of the flow determine the gradient of soil erosion. This is the major factor when it comes to water erosion, and sometimes even in case of glacial erosion.

Climate: One of the major soil erosion causes, climate determines the precipitation levels and wind velocity, which in turn effect soil erosion. More precipitation means more surface flow, and more surface flow means more area vulnerable to erosion by running water.

Similarly, if the wind velocity is high, erosion will also be high and eroded material will be carried farther. The climate factor plays an important role in case of wind erosion and water erosion.

These were some of the major soil erosion causes which are depleting the soil cover of the planet at a faster rate than we can imagine. All the geographical processes occurring on the planet are interrelated, and a slight alteration in one tends to result in a domino effect on ten other process, which are directly or indirectly related to each other.

For instance, if soil cover is depleted vegetation cover will deplete, which will in turn affect the food source for humans. It's high time we understand the geological concept of soil erosion causes and effects of the same on humans, and initiate soil conservation and erosion control measures. We have already induced major hazards such as climate change and global warming on the planet, adding more would only mean adding to our own woes.

Erosion Control Methods

Erosion control is basically the practice of controlling the erosion of soil, by restricting the activity of various agents which contribute to this damaging activity. Right from agriculture to construction, every sector is affected by this natural hazard, and this has prompted people to resort to various erosion control methods, in a bid to curb this activity and the various environmental issues related to it.

Soil Erosion Control Methods

The matter in the top layer of the Earth's surface, typically characterised by its ability to support lifeforms, is known as soil. When this top layer of the Earth's surface is eroded due to various agents of erosion, it is referred to as soil erosion. The most prominent agents of erosion acting upon different types of soil are water (surface flow as well as underground water), wind (most prominent in deserts), waves (restricted to the seashore) and ice (restricted to cold mountainous regions and polar areas). The fact that wind erosion and glacial erosion are restricted to certain zones makes them less important as compared to water erosion and wave erosion. In order to prevent soil erosion, the activity of these agents has to be curbed. There are numerous methods by which this can be done, some of the most popular ones are given below.

Water Erosion Control Methods

Water is by far the most prominent agent of soil erosion. The fact that there are numerous sources of surface flow, right from rainwater to rivers, makes it virtually impossible to stop the activity of water.

Therefore, we need to opt for soil conservation methods as an effective tool to counter water erosion. One of the most effective ways to conserve soil is to plant vegetation. There are various tree species to choose from. Planting different species, including trees, shrubs, creepers etc., is a much wiser option than opting for a single species. As plants start growing, their roots spread in the soil and hold it together. Slopes are most vulnerable to erosion by surface water runoff, and thus, hillside erosion control methods are quite popular in hilly regions. In order to curb erosion on the hillside, the best measure is to plant trees along the slope. Planting creepers is a definite advantage, as they grow horizontally, and thus cover more ground. That, however, doesn't mean you decorate the entire slope with creepers. Bigger trees have deeper roots which hold more amount of soil together, and thus, are quite efficient when it comes to soil conservation.

Beach Erosion Control Methods

A natural stream of freshwater flowing downhill from its source in the mountains to meet an ocean or a lake is known as a river. The river water is confined to a channel or a stream bed.

The rivers are formed when group of springs and streams known as headwaters having their origin in the mountains flow down to form a large stream or springs. The stream bed of a river lies between the banks of a river. The large streams are called a river while the smaller ones are called creeks, brooks, rivulets or tributaries. The rivers form the major component of the water cycle. The water in a river is accumulated from precipitation of ground water and also through the release of stored water in natural reservoirs such as glaciers.

Shoreline erosion by waves is as common as the erosion of sand dunes by wind. When we refer to erosion by waves, it includes erosion of sandy beaches as well as the erosion of rocky structures along the coastline. Use of geotextile tubes filled with sand and stones is one of the most popular measures of controlling shoreline erosion. Similarly, breakwater, which is, a protective structure made from large stones to prevent the shore from being washed away, is also known to be quite effective, and considerably inexpensive.

These were some of the most popular soil erosion control methods which are being used in various parts of the world. If not controlled in time, soil erosion can lead to various problems including loss of fertile soil and natural disasters, like land slides. A slope, which is subjected to continuous erosion, is bound to weaken over the course of time, and eventually give way in the form of a disastrous landslide. In order to avoid such disasters associated with erosion, the need of the hour is to come up with soil erosion prevention methods and implement them at the ground level.

How are Rivers Formed?

Formation of Rivers

Every river in this universe has a point of origin and the gravity plays a significant role in the direction of the flow of a river. In areas where the climate is humid, the point of origin of the rivers is from springs. The points of origination of rivers are the marshes, lakes, and melting glaciers. One of the sources of water that replenish the rivers is either the melting snow or the rainwater. This process is known as the precipitation. Another major source of river water is the rain. When it rains heavily in the hills, the water trickles down the steep slopes and flows onto a riverbed. Initially, the water from the hills flows in an evenly distributed fashion and is called surface run-off. When this water flow travels a certain distance, it begins to flow in parallel rills and also gathers momentum. Soon these parallel rills unite to form a stream. As the rills converge with the stream a brook is formed. This brook flows through a valley. The volume of the water in a brook becomes constant when it gains sufficient volume of groundwater. The brook becomes a river when the water level in the brook increases.

Types of Rivers

The rivers are classified on the basis of the sediments it carries. The sediment carried is controlled by factors such as climate, geology and the stream gradient. Here are a few classifications of the rivers.

Youthful River - A youthful river has a steep gradient and very few tributaries. A youthful river is bound to flow quickly and swiftly. A few examples of youthful river include Trinity River and Brazos in the USA, and Ebro River in Spain.

Mature River - A mature river is less steep and flows slowly compared to the youthful river. There are many tributaries that feed a mature river. The sediment deposit is also less. Examples of mature river include St.Lawrence River, Ohio River and River Thames.

Old River - An old river has a low gradient and is depended on flood plains is known as old river. Some of the world famous old rivers include the Ganges, Nile, and Euphrates.

Rivers have been one of the sources of food, water and transport since pre-historic times. The rivers aid the cultivation of crops by supplying water. Historians claim navigation of rivers date back to the Indus Valley Civilization. Rivers of the world are the major source of fresh water and they sustain their own food chain.

How to Improve Clay Soil

For many people, their garden is their soul. They take great pains to help it flourish with flowers and vegetables. But to create a fabulous garden, you need to know what is the condition of the soil in your garden. Basically, soil can be divided into three basic types-sandy, clayey and loamy. So today our topic of concern will be how to improve clay soil.

Before delving deep into the matter let's know what is clay soil. Soil which is compact, retains moisture even when completely drained and is sticky in nature is known as clay soil. It is often termed as 'heavy soil' and it mostly consists of organic matter, minerals, clay, silt, sand, water and some air. It also retains moisture for a longer period. You can do a simple test in your yard to determine the condition of your soil. Take a handful of the soil. If the soil is sticky and clings to your hand, it is clay soil. Here is another test. For this you will need:

- A quart jar
- Clean water
- A small amount of the soil.

Fill 2/3 of the jar with clean water. Now take the soil, place it inside the jar and close the lid tightly. Give it a vigorous shake for a minute or two. See that the soil has broken down at the bottom of the jar. Allow it to settle down for approximately an hour. The sand layer which is the heaviest will settle at the bottom. The silt layer is at the next level and finally it is the clay layer which will be at the top. The condition of the soil is determined by the percentage of each layer. If the soil has 50 to 100 percent clay then it is clay soil. So, how will you combat the clay soil problem in your yard? Though clay soil is capable of retaining a good amount of moisture and essential nutrients, it may cause a number of problems for the gardener. Firstly, it is difficult for the roots of the plants to penetrate deep inside the soil, resulting in arrested development of the plant. Secondly, clay often contains a high level of alkali which again is hazardous for the plants. Thirdly, it is not easy to till clay soil because it is heavy and sticky. The traditional practice for clay soil treatment is to add gravel to the bottom of the planting hole. But doing so will make the matter worse since the soil above the gravel will retain more water. The process is called 'perched water table'. One of the best options to improve clay soil is to add organic fertilizers like cottonseed meal, fish emulsion, blood meal (blood usually collected from slaughter houses), seaweed fertilizer, sewer sludge and manure though it does not contain much of the nutrients to make it a reliable choice for home gardening. You can also use compost which are basically of three types:

1. Biological: Biological compost is the best and the most reliable. It is often called *black gold* because it provides valuable

nutrients to the plants. Biological compost bags usually have holes so that the microbes that live in it can breathe and breed.
2. Commercial: It is usually made from sewage sludge and debris collected from construction sites. This type of compost is sold in sealed bags and may give out a stale or sour odour.
3. Industrial: It is the least preferred amongst all the composts. It contains high amounts of toxic salts and alkali and is usually derived from chemically burnt saw dust and rice hulls.

Try to mend the soil by mixing it with coarse sand and organic matter. Include a large area rather than a small one since it will limit the plants' growth. Another alternative is to grow trees and shrubs that are best-suited for clay soil e.g. apples, cotton, coffee, willow, black walnut, siberian peashrub, alpine currant, etc. These are some ways in which you can treat and improve clay soil. So, best of luck for your dream garden and let it be one in a million.

Growing Seeds Without Soil

Is soil an indispensable part of seed germination? No. You successfully grow seeds without soil. Read on for more information about growing seeds without soil.

The conventional method of gardening and crop cultivation involved tilling of land, sowing seeds, weeding, fertilizing, etc. Advent of technology revolutionized the traditional methods of gardening and plant propagation. Earlier, it was thought that soil is necessary for the germination of seeds and growth of plants. But, with researches proving the fact that it was not the soil, but the nutrients and minerals in it, which is responsible for the healthy growth of plants, a wide range of mediums and equipment were invented, for growing seeds without soil and for growing house plants without soil. One such example is hydroponic gardening, which uses water as a growing medium for plants, with or without using other growing mediums (perlite, gravel, mineral wool, etc.), except soil. Coming back to growing seeds without soil, there are various methods, which can help you in this task. Let us take a look at the different methods of sprouting seeds without soil.

How to Grow Seeds Indoors Without Soil?

People often come up with doubts like, how do you plant a seed without soil? The following paragraphs deal with the various methods, which can be used for growing seeds without soil. You can choose a method, which is best suited for you and try it out.

Growing Seeds in Paper Towels

It is one of the easiest methods for growing seeds indoors, without using soil. You have to collect materials, like, paper towels, zip pouch bags, a plate, seeds, etc. Take two damp paper towels and spread one of it on the plate. Sprinkle and spread the seeds (so as to form a single layer) on it and cover it with another damp paper towel. Now, slide the plate into a zip pouch bag (unlocked position) and leave it in the light for two to three days. You must not touch it until sprouts are formed. Now, you can transplant the sprouts in pots or even use it for consumption. You may dip the seeds in a diluted bleach solution for at least 15 seconds and then place it on the paper towel without rinsing with water. This step can prevent mold formation in the germination bags. This method is also known as growing seeds in a plastic bag, as the latter is used for growing seeds without soil. Read more on growing apple trees from seed.

Growing Seeds in Water

Water is another medium, which can be used for growing seeds without soil. Take a wide-mouthed jar and add the seeds into it. Add enough water to soak them up and keep it overnight. Next morning, use a cheesecloth to cover the mouth of the jar and fasten the cloth with a rubber band. Strain the water by keeping the jar upside down. Add some more water and shake the jar and drain it once again. Now, you can keep the jar on a plastic tub to drain out excess water. The jar can be rested on the side of the dish, to ensure proper water drainage. Repeat the process for the next two days or until sprouts begin to appear.

There are various other mediums, like, sponge, rock wool, specialised gels (as used in tissue culture), etc., where you can grow seeds without soil. You can try these methods for growing seeds from fruits too. Growing plants without soil is not an impossible task. The above mentioned methods for growing seeds without soil are very easy and less messy and you can also become successful in growing different types of house plants without soil.

Soil Testing

When selecting a commercial laboratory for soil testing, it is generally advisable to choose a local one.

Figure: Soiling Testing

Plants need the right nutrients for their growth and these nutrients are absorbed from the soil in which they grow. Often times though the soil may lack the exact nutrients needed by the plant and so it is necessary to provide these by way of fertilizers. The thing is you need to know the soil composition, so you know exactly how much or how little fertilizer to add. Too much can harm the environment and too little won't have much effect on plant growth. To effectively know how much fertilizer is required, samples of the soil are taken and tested in a laboratory. This is known as soil testing.

Reasons for Testing

Soil Testing is necessary for the following reasons-
- To know the characteristics of the soil.
- To know what fertilizer is required for overcoming deficiencies in the soil.
- To know what quantity of fertilizer is needed so that fertilizers are used only as needed, not too less or too much. This saves money as you don't have to buy more than necessary and also help protect the environment.
- To be able to follow better agricultural practices and achieve higher agricultural production.

Collecting Soil Samples

Here's how to collect soil samples for testing -
- Collect separate samples from each field.
- Collect samples from different areas of the same field and mix them up thoroughly.
- If the same field show has distinct areas with different soil appearances, take different samples and keep them separate.
- Make a map of your field or fields and mark the areas from which the samples were taken.
- Dig to about plow depth to take samples.
- Take samples with clean tools. If the soil is soft, you can easily take samples using a trowel, spade or soil tube. If the ground is hard, you may have to use an adze, screw auger or post hole auger.
- Clean tools before taking sample from another area.
- Collect the samples in a clean bucket.
- Clean bucket if it has been used before to take samples from another field or if it going to be used again. This is to avoid any soil mix or contamination.
- Pack the soil samples carefully.
- Mark the samples so you don't mix them up.

- Get soil sample information forms.
- Fill them out for each soil sample.
- Provide correct information about land topography, irrigation, drainage, crop varieties grown currently and in the past.
- Send filled forms together with the soil samples to the laboratory for soil testing.

Testing Method

When selecting a commercial laboratory for soil testing, it is generally advisable to choose one that is local for the following reasons-

- The people at the local laboratory are likely to know more about local garden and farming conditions.
- They will have more experience in analysing the local soil.

Soil testing needs to be done as soon as possible, preferably within 24 hours.

Or there is a chance of chemical changes taking place within the soil. If testing can't be done at once, then the soil must be frozen or air dried to make it stable for a longer period.

There are various kinds of tests that can be carried out. You need to have specified in your forms which tests are required for your purpose.

Here are some tests that are carried out to test soils -

- Testing for the major nutrients - Nitrogen, Phosphorus and Potassium
- Testing for the secondary nutrients - Sulfur, Calcium and Magnesium
- Testing for the minor nutrients - Iron, Manganese, Copper, Zinc, Boron, Molybdenum and Aluminium
- Testing for Soil Acidity
- Testing for the Electrical Conductivity in the soil.
- Testing for the organic matter in the soil
- Testing for the moisture in the soil
- Testing for contamination, if any, in the soil.

If you don't require quite such comprehensive tests and don't want to go through all the formalities of sending samples to the laboratory, you can carry out some tests yourself with a do-it-yourself testing kit. You can buy these from the lab or from an agricultural products store. The number of tests possible with these kits is quite limited, but you should be able to do the following -

- Testing for the major nutrients - Nitrogen (N), Phosphorus (P) and Potassium (K)
- Testing for Soil Acidity.

But keep in mind that the lab report will be more detailed and probably more accurate.

Granite Rock Facts

These interesting granite rock facts have played an important role in soaring popularity of granite rocks. In order to know more on these facts about granite rocks, read on...

Granite is an igneous rock, characterised by a hard coarse grained surface, found in abundance on this planet. It contains several minerals, most prominent ones being quartz and feldspars, all of which are locked into each other, making granite immensely strong as well as durable. Hence is the first choice in the field of construction. Other than its strength, there are several granite rock facts, which make it one of the most fascinating object found on our planet.

How is Granite Formed?

Post a volcanic eruption, the molten rock or magma seeps in between the other rocks in the Earth's crust. The magma, referred to as lava when in the interior of the Earth, is formed due to melting of various types of metamorphic rocks. As it cools, the magma tends to crystallize owing to its high silicate content and leads to formation of granite rock. The entire process of granite rock formation is known to take millions of years.

Granite Etymology

One of the most interesting facts about granite rocks is its etymology. The word granite is derived from the Latin word - *granum*, meaning grains. The rock is referred to as granite owing to its coarse grained structure.

Granite Rock: Types of Igneous Rocks

Although basalt and granite rocks are types of igneous rock they have a drastic difference in composition. An interesting piece of granite rock information is that some granite rocks existing today have been there since the geologic periods, which makes them the oldest rocks on the planet.

Granite Rock Composition and Density

Silicon dioxide and aluminium oxide together constitute more than 85 percent of a granite rock composition. Other chemicals include sodium oxide, potassium oxide, etc. Its average density is 2.75 g/cm. The molecular structure of the rock makes it very rigid and stable. One of the most interesting granite rock facts is its colour. The colour, which varies from pink to various shades of black, is largely dependent on the presence of minerals in it.

Granite Stone Occurrence

Granite is found in abundance, all over the planet, but it is only restricted to Earth's crust, where it had seeped in during the geologic periods. It is assumed to be the most abundant basement rock, lying beneath a thin layer of sedimentary rocks. Owing to its presence in bulk, commercial granite harvesting is a flourishing business in various countries of the world including India, Brazil and United States.

Other Interesting Granite Rock Facts

When compared with marble, granite is much more harder and hence less vulnerable to flaw lines and pitting. Granite, like most of the other natural stones, is radioactive indeed. Although the uranium count of granite rocks, exceeds that of several other natural stones, it doesn't pose any health risk to humans coming into contact. Read more on radioactivity.

Granite Rock Uses

Being strong and durable, granite rock has been used for construction since several centuries. But its use is not just restricted to construction. It also sports superb bacteria resistance qualities, second only to stainless steel. Granite rocks, when polished, become scratch proof and stain resistant. Being easy to clean and heat resistance, granite rock is a popular choice for kitchen countertops. Other than construction and interiors, granite rock is also used as to create sculptures (Mount Rushmore being the best example). It is also a delight for amateur as well as professional rock climbers. Read more on granite countertops.

These were some of the most prominent granite rock facts which support the claim that granite is one of the most strong, durable, abundantly found and vastly used rock in the world. It doesn't just add to the durability of the structure, but also contributes to its versatility and grace. No wonder, it's the first choice for many people across the world when it comes to building a house or decorating the interiors.

Acidic Soil Plants

The following article will give you brief information about acidic soil plants, in a bid to explain why these plants grow well in acidic soil. Continue reading for basic information about certain species of plants which require acidic soil for proper growth.

Certain species of plants grow well in certain types of soil, and you don't have to be some veteran in the field of gardening to know that. Basically, soil composition has a crucial role to play when it comes to plant growth, and when we talk about the composition of soil its pH value is one crucial factor which has to be taken into consideration. While most of the plant species grow well in soil with a pH balance of 6 - 7, others require soil with a pH balance much less than that - ideally 4 - 6, in order to facilitate proper growth. Soil with low pH is known as acidic soil, while the plants which grow well in such soil are known as acidic soil plants or ericaceous plants.

Acidic Soil

When it comes to soil science, the different types of soil are broadly categorised into two groups - acidic soil and alkaline soil, on the basis of their pH balance. Soil with a pH balance of 7 or less is considered to be acidic soil, while soil with a pH balance exceeding 7 is known as alkaline soil.

Similarly, a value of 7 on the pH scale represents neutral soil. Acidic soil is predominantly formed as a result of decomposition of plant and animal matter in this soil. Other than this, soil formed from an acidic parent rock will also turn out to be acidic soil. Generally, you come across acidic soil in regions which have soft water or which experience heavy rainfall. Acidic soil can also be transformed to alkaline soil as a result of various natural occurrences, including disintegration of a limestone rock.

Acidic Soil Plants

The simple rule of the thumb is that the plants which prefer acidic conditions grow well in acidic soil. This is so because soil pH levels determine the presence of different nutrients in this soil. If you try to plant these species of plants in alkaline soil, it will indirectly affect their growth as they will not be able to derive the necessary nutrients from the soil. As in case of these plants that love acidic soils, several microorganisms present in the soil and diseases affecting these plants and microorganisms have a preference for either acidic or alkaline soil. That being said, the pH levels of soil have a crucial role to play when it comes to smooth functioning of the said ecosystem.

Best Plants for Acid Soil

When we talk about plants for acidic soil, the first names that are likely to come to your mind would be rhododendron and heather. While these are the most popular acidic plant species, there also exist several more species which prefer to grow in acidic soils with a pH value of less than 6. Given below are some examples of acidic soil plants on the basis of their genre.

- Some popular species of trees which prefer acidic soil are Strawberry Tree, Sweetbay, American Sweet Gum, Koster Blue Spruce, etc.
- Some species of perennial plants which are known to grow well in acid soils include the Tibetan Blue Poppy, Flag Iris, Big Blue Lily-turf, etc.
- Some shrubs and bushes which are often associated with acidic soils include Winter Flowering Heath, Bayberry, Ground Juniper, Chinese Witch Hazel, Calico Bush, etc.
- When climbers which prefer acidic soils are taken into consideration, the most prominent names include the Trumpet Creeper and the Persian Ivy.
- If plant species that work as a ground cover is something that you are looking forward to, the Heather and the Wintergreen species are the most popular.

Some of these acidic soil plants have beautiful flowers, and this is what makes them quite popular among gardening enthusiasts. If you are one of these people looking forward to add some acidic soil loving plants to your garden, you should try some of the species given above. If your region doesn't have acidic soil or if its pH levels are high, you can resort to various methods of lowering soil pH including some simple ones such as adding sulfur or composting. Similarly, if the soil in the area wherein you reside is too acidic for the plants to thrive, you can resort to neutralisation of soil by adding agricultural lime or calcium sulfate to it.

Types of Landforms

Landforms are defined as the natural physical features found on the surface of the earth. Landforms are created as a result of the various forces of nature such as wind, water and ice and also by the movement of the earth's tectonic plates.

For example, due to these actions, the soil gets eroded and deposited somewhere far from the site of erosion, thus leads to the formation of different landforms. Some landforms are created in a matter of few hours; others take millions of years to appear. A group of landforms in a particular area is called its landscape.

Types and Characteristic Features of Landforms

There are many types of landforms on the earth's surface. Each landform is characterised by its slope, elevation, soil and rock type, stratification and orientation. Following is the list of some of the common types of landforms and their characteristics:

Mountains: Mountains are areas, which are higher than the surrounding areas and are characterised by a peak, e.g. The Himalayas. Surprisingly, they are more frequently present in the oceans than in land. A mountain is steeper than a hill. Mountains are formed due the tectonic movement such as an earthquake or a volcanic eruption. A few are resulted due to erosion of the surrounding areas by the action of wind, water or ice.

Plateaus: Plateaus are large highland flat areas separated from the surrounding areas by a steep slope, e.g. The Tibetan plateau. Plateaus are formed due to various actions such as collision of the earth's tectonic plates, uplift of the earth's crust by the action of magma; some are resulted due to the lava flow from the volcanic eruption.

Islands: Islands are areas that are completely surrounded by water, e.g. The Hawaiian Islands. Islands are formed either as a result of the volcanic eruption or due the presence of hot spots on the lithosphere.

Plains: Plains are flat areas or low relief areas on the earth's surface, e.g. prairies, steppes. Plains are formed due to the sedimentation of the eroded soil from the hills and mountains or due to the flowing lava deposited by the agents of wind, water and ice.

Valleys: Valleys are flat areas of land between the hills or mountains, e.g. The California Central Valley. Mostly they are formed by the actions of rivers and glaciers. Depending upon the shape, valley forms are classified as U-shaped or V-shaped valley. V-shaped valleys are formed by flowing water or rivers, whereas U-shaped valleys are formed by glaciers.

Deserts: Deserts are very dry lands with little or no rainfall, for example, The Sahara desert. Mostly deserts are formed in rainshadow areas, which are leeward of a mountain range with respect to the wind direction. Thus, the mountains block the passage of wind resulting in little or sometimes no rain.

Loess: Loess are deposits of silt and with a little amount of sand and clay. Many times wind action is responsible for formation of loess; however sometimes glacial activity can also form loess.

Rivers: Rivers are natural flowing stream of freshwater, e.g. The Nile. They mostly flow towards lakes or oceans but sometimes they dried up without reaching another water body. River water is collected from the surface water runoffs, groundwater water recharge and sometimes from the water reservoirs such as glaciers. Landforms definitely play an important role in the formation of rivers.

Oceans: Oceans are the biggest form of water and are saline, such as The Pacific Ocean. Oceans of the world cover around 71% of the earth's surface and control the weather and climate of the earth's surface. Oceans are originated due the Continental Drift, i.e. the movement of the earth's tectonic plates.

Glaciers: Glaciers are huge slow moving body of ice. Glaciers are formed due to the compaction of snow layers and moves with respect to gravity and pressure. Mainly there are two types of glaciers - Alpine glaciers, which are formed in high mountains and Continental glaciers, which are formed in cold Polar Regions.

Rock: Types of Rocks

The lithosphere is defined as "the solid part of the earth consisting of the crust and outer mantle" and this very part is what homes all the different rock types. Generally, there are 3 types of rocks, namely, igneous, sedimentary and metamorphic. The reason why rocks are classified under three different forms comes from their mineral and chemical composition. The classification is also governed by the texture which the constituent particles of the rocks are made of and the different processes which are responsible for forming them. Now let's go into these classification of rock types and furthermore into their sub types. Read more on identifying rocks.

Different Rock Types

Igneous Rocks

The first one coming up in the list of rock types is the igneous rock type. These rocks have their formation credited to the cooling of magma or lava. These rocks may get formed below the surface of the earth or on the surface of the earth. Rocks forming below the surface are known as intrusive or plutonic igneous rocks and those which are formed below the surface, are known to be as extrusive igneous rocks. Apart from these two types of igneous rocks, there are other kinds which are rare or let's say, less common. These rocks are formed at a depth in between the plutonic and volcanic rocks. These rocks are known as hypabyssal igneous rocks.

The magma, under the ground gets trapped in small pockets. Then a gradual cooling process begins, which results in a transformation of the trapped magma into what is known as igneous rocks. As the cooling process is slow, these rock types are coarse grained. Coming to the extrusive rocks, they are formed by the magma which rises to the surface of the earth. The cooling and solidification process takes place rapidly and that is the reason why these rock types are fine grained. Common examples of igneous rocks are basalt (an extrusive rock), granite (an intrusive rock) and andesite (a hypabyssal rock).

Sedimentary Rocks

These types of rocks may be attributed as secondary rocks, as they would not have been named so if igneous rocks would not have been present in the first place. Sedimentary rocks, as the name suggests, are formed by years of deposition of layers upon layers of sediments, which may consist of organic, chemical or mineral origin. The deposition is followed by compaction of the particulate matter and cementation and finally the formation of the sedimentary rocks. These rock types can take birth either in water or on the land surface. Read more on minerals and rocks for kids. There are three types of sedimentary rocks and their characteristic feature depend upon the process which created them.

They include the clastic, chemical precipitate and biochemical sedimentary rock types. The first kind, as the name suggests, comprises casts. These are derived from other minerals and other weathered igneous or metamorphic rocks.

Chemical precipitate sedimentary rocks are formed due to the action of water of dissolving several minerals and having them deposited on evaporation. Biochemical sedimentary rock types, are named according to their formation which occurs as a result of accumulation or deposition of remnants of various organisms.

Typical examples of sedimentary rocks include sandstone (a clastic rock), gypsum (a chemical precipitate rock) and stromatolite (a biochemical sedimentary rock).

Get elaborate explanation on this subject by reading more on:
- Sedimentary Rock Formation
- Sedimentary Rock Facts.

Metamorphic Rocks

Igneous and sedimentary rock types may get subjected to humongous intensity of heat or pressure so that they undergo a complete transformation. They become what is known as metamorphic rocks. The terms 'meta' means change and 'morph' signifies form. These rocks are formed deep within the Earth's crust, where the minerals of the buried igneous and sedimentary rocks become unstable and out of equilibrium with their new environmental conditions. The types of metamorphic rocks are the ones which are formed by contact metamorphism or regional metamorphism. The first phenomenon occurs as a result of magma being injected into the surrounding solid rock, which creates the morphism. The latter one is a natural process, wherein rocks which lay at great depths below the Earth's surface and get subjected to intense pressure caused by the immense weight of the rock layers above, get metamorphosed. Common examples include amphibolite, granulite, hornfels, marble, etc. Learn more on metamorphic rock facts: types of metamorphic rocks.

The study about rocks and the different rock types (known as petrology) has much more than what has been outlined in this brief discussion. Merely reading about such important fragments of the

Earth in black and white, may not fulfill your endeavour of getting an in-depth grasp about nature and its wonders. So, not only know them theoretically, check them out for real!

Moisture Metre: Moisture Detection and Analysis

The moisture metre, as the name suggests, is used for moisture detection and moisture analysis. This is particularly used for the detection of water percentage in wood, plants, soil, concrete etc. There are various companies which manufacture moisture metres and they are widely used. Let us take a deeper look at understanding the various aspects of a moisture metre and its uses.

Figure: Manufacturer of moisture measuring technology for bulk materials

You must have noticed moisture metres in the hands of people who are into wood work or carpentry.

They need it for moisture detection and moisture analysis of wood, to make out if the wood they have selected will suit their purpose or not. However, the use of moisture metres is not limited to moisture detection and moisture analysis in wood only. Let us take a look at the different types of moisture metres that are used for various purposes.

Moisture Metre for Wood

The wood moisture metre works on the principle that the wood resistance decreases, as the moisture content in the wood increases. The decrease in the resistance is converted into digital data and the moisture metre displays the percentage of water in the wood. The moisture metre consists of two electrodes, which are inserted into the wood, to find out the resistance between them and hence its moisture content.

For ideal wood work, the amount of moisture in the wood should be at equilibrium with respect to its surroundings. If the moisture content in the wood changes, it may result in a change of the shape of the wood. You must have noticed that when a piece of wood is dried for a long time, it loses its shape. People who are into carpentry, want to avoid this and hence, a moisture metre serves their purpose of moisture detection and moisture analysis.

Moisture Metre for Concrete

Concrete moisture metres were introduced to check out the moisture content in the concrete slabs that are used in construction. Their principle of working and finding out the moisture content, is same as that of the wood moisture metre.

The resistance between their electrodes is checked out here too, to find out the moisture content in the concrete. However, they significantly differ in their moisture detection and moisture analysis methods, from the wood moisture metres. They can measure moisture at a depth of 1 inch from the moisture surface. This is required because people working in the field need to know the moisture content at the depths of the concrete slab, to analyse the construction.

Moisture Metre for Soil

There are companies which manufacture moisture metres for measuring soil moisture levels too. This moisture analysis is required to find out if the plants in your garden would require watering or not. It may happen that because of rain, the moisture content in the soil is already sufficient and your plants do not need any more water, for a considerable amount of time. Under such cases, a soil moisture metre can be used for the moisture detection and moisture analysis in the soil, to find out regularly, if the plants require any additional watering. If it rains regularly and your plants are getting over watered, then the moisture analysis can help you in determining if you need to take some preventive measures, to cover up your plants from the rain. The soil moisture metre works on the same principle as that of any other ordinary moisture metres.

When using a moisture metre, you should follow the manufacturer's instructions or the manual. Always buy a moisture metre that can easily fit into your grip. Also, check that the electrodes on your moisture metre are flexible and bendable. You can easily carry around a wood moisture metre or concrete moisture metre along with you, while a soil moisture metre needs to be put in the soil of the planted area. If your plants need watering, then it will be indicated on the moisture metre. It is one of the most handy devices and solves the most important purposes of moisture detection and moisture analysis, not only in the wood work and construction business, but also for our gardens.

Chapter 5
Soil Profile

Soil Profile refers to the layers of soil; horizon A, B, and C. If you're wondering what horizon A is, here's your answer: horizon A refers to the upper layer of soil, nearest the surface. It is commonly known as topsoil. In the woods or other areas that have not been plowed or tilled, this layer would probably include organic litter, such as fallen leaves and twigs. The litter helps prevent erosion, holds moisture, and decays to form a very rich soil known as humus. Horizon A provides plants with nutrients they need for a great life.

The layer below horizon A, of course, has to be horizon B. Litter is not present in horizon B and therefore there is much less humus. Horizon B does contain some elements from horizon A because of the process of leaching. Leaching resembles what happens in a coffee pot as the water drips through the coffee grounds. Leaching may also bring some minerals from horizon B down to horizon C. If horizon B is below horizon A, then horizon C must be below horizon B. Horizon C consists mostly of weatherized big rocks. This solid rock, as you discovered in Soil Formation, gave rise to the horizons above it.

Soil Profile

Sand, silt, and clay are the basic types of soil. Most soils are made up of a combination of the three. The texture of the soil, how it looks and feels, depends upon the amount of each one in that particular soil. The type of soil varies from place to place on our planet and can even vary from one place to another in your own backyard.

Soil erosion, caused by wind and rain, can change land by wearing down mountains, creating valleys, making rivers appear and disappear. It is a slow and gradual process that takes thousands, even millions of years. But erosion may be speeded up greatly by human activities such as farming and mining. Soil develops very slowly over a long period of time but can be lost too quickly. The clearing of land for farming, residential, and commercial use can quickly destroy soil. It speeds up the process of erosion by leaving soil exposed and also prevents development of new soil by removing the plants and animals that help build humus.

Today's farmers try to farm in a way that reduces the amount of erosion and soil loss. They may plant cover crops or use a no-till method of farming. Soil is an important resource that we all must protect. Without soil there is no life.

Organic Matter

Organic matter (or organic material, Natural Organic Matter, or NOM) is matter that has come from a once-living organism; is capable of decay, or the product of decay; or is composed of organic compounds. The definition of organic matter varies upon the subject for which it is being used.

Organic matter is broken down organic matter that comes from plants and animals in the environment. Organic matter is a collective term, assigned to the realm of all of this broken down organic matter. Basic structures are created from cellulose, tannin, cutin, and lignin, along with other various proteins, lipids, and sugars. It is very important in the movement of nutrients in the environment and plays a role in water retention on the surface of the planet. These two processes help to ensure the continuance of life on Earth.

How Organic Matter is Created

All living and growing matter on this planet contains organic components. Different types of matter include humans, animals, plants, and microorganisms. After the living matter dies, it decomposes. The organic matter from them and their excretions is broken down through an unknown reactive process into natural organic matter. Larger molecules of organic matter can be formed from the polymerization of different parts of already broken down matter. The relative size, shape, and composition of a molecule of organic matter is very random. "NOM can vary greatly, depending on its origin, transformation mode, age, and existing environment, thus its bio-physico chemical functions and cheese vary with different environments."

Natural Ecosystem Functions

Natural organic matter is present throughout the ecosystem. After degrading and reacting, it can then move into soil and mainstream water via waterflow. NOM forms molecules that contain nutrients as it passes through soil and water. It provides nutrition to living plant and animal species. NOM acts as a buffer, when in aqueous solution, to maintain a less acidic pH in the environment. Little is known why this occurs but research shows the buffer acting component to be crucial to wean away the cheese of acid rain.

Source Cycle

A majority of NOM not already in the soil comes from groundwater, which is water under the surface of the earth. When the groundwater saturates the soil or sediment around it, NOM can freely move between the phases. But, the groundwater has its own sources of natural organic matter too:

- "organic matter deposits, such as kerogen and coal
- soil and sediment organic matter
- organic matter infiltrating into the subsurface from rivers, lakes, and marine systems"

Note that one source of groundwater is soil organic matter and sedimentary organic matter. The major method of movement into soil is from groundwater, but NOM from soil moves into groundwater as well. Most of the NOM in lakes, rivers, and surfaced water areas comes from deteriorated material in the water and surrounding shores. However, NOM can pass into or out of water to soil and sediment in the same respect as with the soil.

Importance of the Cycle

Natural organic matter uses all these different phases (soil, sediment, water and groundwater) to move throughout the environment. This action of movement creates a cycle. Things decompose into NOM, travel through water flow or soil, and then are free to spread through the phases. If it were not for this cycle, important nutrients such as minerals, vitamins, and metals would not be as easily spread throughout the surface of the Earth. Furthermore, this shows there are no independent processes in the environment, which means everything is connected in some regard. Physical, biological, and chemical systems work together to create natural processes.

Soil Organic Matter

The organic matter in soil derives from plants and animals. In a forest, for example, leaf litter and woody material falls to the forest floor. This is sometimes referred to as organic material. When it decays to the point in which it is no longer recognisable it is called soil organic matter. When the organic matter has broken down into a stable substances that resist further decomposition it is called humus. Thus soil organic matter comprises all of the organic matter in the soil exclusive of the material that has not decayed.

One of the advantages of humus is that it is able to withhold water and nutrients, therefore giving plants the capacity for growth. Another advantage of humus is that it helps the soil to stick together which allows nematodes, or microscopic bacteria, to easily decay the nutrients in the soil.

There are several ways to quickly increase the amount of humus. Combining compost, plant or animal materials/waste, or green manure with soil will increase the amount of humus in the soil.

1. Compost: Decomposed organic material.
2. Plant and animal material and waste: Dead plants or plant waste such as leaves or bush and tree trimmings, or animal manure.
3. Green manure: Plants or plant material that is grown for the sole purpose of being incorporated with soil.

These three materials supply nematodes and bacteria with nutrients for them to thrive and produce more humus, which will give plants enough nutrients to survive and grow.

Decay

Organic matter may be defined as material that is capable of decay, or the product of decay (humus), or both. Usually the matter will be the remains of recently living organisms, and may also include still-living organisms. Polymers and plastics, although they may be organic compounds, are usually not considered organic material, due to their poor ability to decompose. A clam's shell, while biotic, would not be considered organic matter by this definition because of its inability to decay.

Organic Chemistry

Measurements of organic matter generally measure only organic compounds or carbon, and so are only an approximation of the level of once-living or decomposed matter. Some definitions of organic matter likewise only consider "organic matter" to refer to only the carbon content, or organic compounds, and do not consider the origins or decomposition of the matter. In this sense, not all organic compounds are created by living organisms, and living organisms do not only leave behind organic material. A clam's shell, for example, while biotic, does not contain much organic carbon, so may not be considered organic matter in this sense. Conversely, urea is one of many organic compounds that can be synthesized without any biological activity.

Very little is currently known about natural organic material. Scientists are unable to crystallize it. This is important because once you can crystallize the material, it can be isolated and studied with x-ray crystallography. This method is standard for determining unknown compounds. NOM has not been characterised either and no unique structure is known. The best way to characterise NOM is by discovering chemical, physical, and thermodynamic properties of the matter. Analytical techniques are currently being discovered to allow this to happen. The only information scientists have is that NOM is heterogeneous and very complex. Generally, NOM, in terms of weight, is:

- 45-55% Carbon
- 35-45% Oxygen
- 3-5% Hydrogen
- 1-4% Nitrogen.

The molecular weights of these compounds can vary drastically, depending on if they repolymerize or not, from 200-20,000 amu(4). It is also important to know that 10-35% of the carbon present forms aromatic rings. These rings are very stable due to resonance stabilisation, so they are difficult to break down. The aromatic rings are also susceptible to electrophilic and nucleophilic attack from other electron-donating or electron-accepting material, which explains the possible polymerization to create larger molecules of NOM.

There are also reactions that occur with NOM and other material in the soil to create compounds never seen before. Unfortunately, it is very difficult to characterise these because so little is known about natural organic matter in the first place. Research is currently being done to figure out more about these new compounds and how many of them are being formed.

Organic Matter in Water

Water Purification

The same capability of natural organic matter that helped with water retention in soil creates problems for current water purification methods. In water, NOM can still bind to metal ions and minerals. These bound molecules are not necessarily stopped by the purification process, but do not cause harm to any humans, animals, or plants. However, because of the high level of reactivity of natural organic matter, byproducts that do not contain nutrients can be made. These byproducts are much larger and can induce biofouling, which essentially breaks down water filtration systems in water purification facilities. The larger molecules clog the water purification filters intended to keep material like that out of drinking water. The fact that these byproducts are removed through purification is very good news, but having to replace filters constantly to maintain effectiveness is costly for water treatment businesses. This byproduct problem could be treated by the disinfection technique known as chlorination, which often breaks down residual material clogging systems, but research has shown that the natural organic matter also forms byproducts with this method.

Potential Solutions

A large breakthrough could be underway after a paper published in the Applied and Environmental Microbiology journal showed proof that water with natural organic matter could be disinfected with ozone-initiated radical reactions. The ozone (three oxygens) has very strong oxidation characteristics. It can form hydroxyl radicals (OH) when it decomposes, which will react with the natural organic matter to shut down the problem of biofouling. The article did this on a very small scale of water and natural organic matter, so further research is being done to scale up the reaction. This journal article does show that solutions are on the horizon to prevent our water purification systems from being broken down by NOM.

False Positives

Many water quality groups, such as the North Carolina State University Water Quality Group, believe that having too much natural organic material will cause deoxygenation and essentially remove oxygen from the water. Although organic material, which consists of many hydrocarbon and cyclic carbon chains, is susceptible to attack by oxygen, it would be sterically unfavourable to attach oxygens to every single carbon. Basically, molecules do not enjoy other molecules being too close to them when they have the same electronegativity. Most of this is because of electrostatic charge, which says that opposite charges are attracted and like charges are repelled. If you have oxygens (with a negative charge in theory) bonded to carbons next to each other, they will want to be as far away from each other as possible. Also, a larger molecule like oxygen (relative to carbon) does not want to attach to a carbon that already has oxygens on it when it could attach to a carbon without oxygens on it.

Of course, there are exceptions, such as varying the temperature at which these reactions occur. As the temperature becomes much higher, there is a better chance that an unfavourable reaction will occur because molecules move around faster increasing the randomness of the system (entropy). Yet, as we consider the cold water in the natural environment, it is logical to see that all the oxygen in the water will not be consumed by NOM.

Estimate of Oxygen in Surface Water

This can be proved numerically. The approximate volume of liquid water on the Earth is $1.333*10^9$ km^3, or $1.333*10^{24}$ millilitres. Since the density of water is roughly 1g/mL, this is $1.333*10^{24}$ grams of water. Dividing this number by 18.02 grams per mole gives $7.399*10^{22}$ moles of water. Each mole of water has one mole of oxygen and each mole of oxygen has 6.022×10^{23} oxygen molecules. Multiplying $7.399*10^{22}$ moles by $6.022*10^{23}$ molecules/mole gives $4.456*10^{46}$ oxygen molecules in the liquid water on the surface of the planet. (Please note this is an estimate taken from average data.)

NOM is not going to use up all the oxygen on the earth and remove water.

Vitalism

The equation of "organic" with living organisms comes from the now-abandoned idea of vitalism that attributed a special force to life that alone could create organic substances. This idea was first questioned after the artificial synthesis of urea by Friedrich Wöhler in 1828.

Soil Organic Matter

Soil degradation has become a major concern in Canada. Erosion, salinization, acidification and loss of organic matter are the main forms of soil deterioration. This publication deals with the role of organic matter in soil productivity and the effects of various management practices on soil organic matter.

What is Organic Matter?

Soil organic matter consists of a variety of components. These include, in varying proportions and many Intermediate stages:

- raw plant residues and microorganisms (1 to 10 per cent)
- "active" organic traction (10 to 40 per cent)
- resistant or stable organic matter (40 to 60 per cent) also referred to as humus.

Raw plant residues, on the surface, help reduce surface wind speed and water runoff. Removal, incorporation or burning of residues predisposes the soil to serious erosion.

The "active" and some of the resistant soil organic components, together with microorganisms (especially fungi) are involved in binding small soil particles into larger aggregates. Aggregation is important for good soil structure, aeration, water infiltration and resistance to erosion and crusting.

The resistant or stable fraction of soil organic matter contributes mainly to nutrient holding capacity (cation exchange capacity) and soil colour. This fraction of organic matter decomposes very slowly and therefore has less influence on soil fertility than the "active" organic fraction.

Organic matter in soil serves several functions. From a practical agricultural standpoint, it is important for two main reasons. First as a "revolving nutrient bank account"; and second, as an agent to improve soil structure, maintain tilth, and minimise erosion.

As a revolving nutrient bank account, organic matter serves two main functions:
- Since soil organic matter is derived mainly from plant residues, it contains all of the essential plant nutrients. Accumulated organic matter, therefore, is a storehouse of plant nutrients. Upon decomposition, the nutrients are released in a plant-available form.
- The stable organic fraction (humus) adsorbs and holds nutrients in a plant available form.

Organic matter does not add any "new' plant nutrients but releases nutrients in a plant available form through the process of decomposition. In order to maintain this nutrient cycling system, the rate of addition from crop residues and manure must equal the rate of decomposition.

If the rate of addition is less than the rate of decomposition, soil organic matter will decline and, conversely if the rate of addition is greater than the rate of decomposition, soil organic matter will increase. The term steady state has been used to describe a condition where the rate of addition is equal to the rate of decomposition.

Fertilizer can contribute to the maintenance of this revolving nutrient bank account by increasing crop yields and consequently the amount of residues returned to the soil.

Organic Matter in Virgin and Cultivated Soils

Soils in Alberta are divided into soil groups (zones) based on the amount of organic matter they contain. They occur in geographic zones from the southeast to the northwest and are identified as the Brown, Dark Brown, and Black Chernozemic (prairie) soils.

The Brown soils have the least amount of organic matter because of the relatively small inputs of plant residues contributed by the short grass prairie vegetation under which these soils developed. Black soils developed under cooler and wetter conditions which allowed for more grass growth and thus a greater accumulation of organic matter.

Further north and west, trees became the dominant vegetation. Soils influenced by forest vegetation for a moderate length of time constitute the Dark Gray or transitional soils. Where the forest cover was established for a longer period, Luvisolic (forest) soils developed. Organic (peat) soils occur in low lying areas throughout the Black, Dark Gray and Gray soil zones. These soils are saturated with water for much or all of the year thereby reducing the rate of organic matter decomposition.

The amount of soil organic matter characteristic of virgin and cultivated soils in the various zones is shown in Table 1. Cultivation generally has resulted in a 30 to 50 per cent loss of organic matter.

Table 1. Organic matter in native and cultivated soils (per cent)

Soil zone	Virgin	Cultivated
Brown	3-4	2-3
Dark Brown	4-5	3-4
Black	6-10	4-6
Dark Gray	4-5	2-3
Gray	1-2	1-2

Before our soils were cultivated, they had achieved a "steady state". In most of our prairie soils, the increased rate of decomposition associated with cultivation, combined with the low rates of crop residue addition associated with crop-fallow rotations has caused a fairly rapid decline in soil organic matter. The rate of decline decreases with time as the amount of total soil organic matter decreases and particularly as the "active" organic fraction is depleted.

Cultivation of soils that are naturally high in organic matter will usually result in a decrease of organic matter. In the case of Luvisolic soils, their poor physical properties and low fertility have encouraged the use of forages, fertilizers, manure and judicious tillage. Such management practices have resulted in an increase in soil organic matter on Luvisolic soils, whereas excessive tillage, fallowing and minimal fertilization have lead to further depletion of the soil organic matter.

Effects of Organic Matter Decline

As stated in the introduction, soil degradation is becoming a major concern in Canada. Loss of organic matter is often identified as one of the main factors contributing to declining soil productivity, but it is misleading to equate a loss in soil organic matter with a loss in soil productivity.

Soil organic matter contributes to soil productivity in several ways, but there is no direct quantitative relationship between soil productivity and total soil organic matter. In fact, it has been the decline in organic matter that has contributed to the productivity of the crop-fallow system.

This decline in organic matter has resulted in the release of large amounts of plant nutrients, particularly nitrogen. For example, a decrease in soil organic matter of 2 per cent releases about 2,400 lb/ac of nitrogen. If this decline occurred over a 60 year period, an average of 40 lb/ac/yr of plant-available nitrogen has come from the soil organic matter.

We therefore view prairie soils which had relatively high levels of organic matter as being nitrogen fertile, but this fertility could only be attained under a management system that allowed for organic matter to decline. Frequent fallowing has been a major factor contributing to this decline.

Insofar as organic matter contributes to improved soil physical properties (e.g., tilth, aggregation, moisture holding capacity and resistance to erosion) increasing soil organic matter will generally result in increased soil productivity. But on many soils, suitable soil physical properties occur at relatively low levels of organic matter (2-4 per cent).

A level of organic matter higher than required to produce suitable physical properties is beneficial in that the soil has a greater buffering and nutrient holding capacity, but it does not contribute directly to soil productivity. If soils are managed so organic matter is not declining (steady-state), soils higher in organic matter (e.g., 8 per cent) are not inherently more productive or fertile than those that have less organic matter (e.g., 5 per cent).

To equate the ability to supply nutrients with total soil organic matter is not valid. The "active" fraction of organic matter is a more reliable indicator of soil fertility than is total soil organic matter. In cultivated soil, the "active" fraction is influence mainly by previous management.

Soil organic matter cannot be increased quickly even when management practices that conserve soil organic matter are adopted. The increased addition of organic matter associated with continuous cropping, and the production of higher crop yields, are accompanied by an increase in the rate of decomposition. Moreover, only a small fraction of crop residues added to soil remains as soil organic matter.

After an extended period of time, the return of all crop residues and the use of forages in rotations with cereals and oilseeds may significantly increase soil organic matter, particularly, the "active" fraction.

Managing Soil Organic Matter

There have been vast changes in the nature of agricultural production. In the past, farms were small, and much of what was produced was consumed on the farm. This system allowed for the limited removal of soil nutrients since there was an opportunity to return most of the nutrients back to the land.

The advent of the internal combustion engine, migration from rural to urban communities, increasing farm size and specialisation in production have resulted in a system of production where there is greater removal of plant nutrients from the soil and less opportunity for nutrient cycling.

Maintenance of organic matter for the sake of maintenance alone is not a practical approach to farming. It is more realistic to use a management system that will give sustained profitable production. The greatest source of soil organic matter is the residue contributed by current crops. Consequently, crop yield and type, method of handling residues and frequency of fallow are all important factors. Ultimately, soil organic matter must be maintained at a level necessary to maintain soil tilth. The effects of specific management practices are discussed below.

Summer Fallow

Summer fallowing accelerates the loss of organic matter. Aeration of the soil associated with tillage, and the increase in soil temperature and moisture results in increased organic matter decomposition. Since little In the way of residues are added to the soil, a net loss of organic matter occurs. Research has shown that as the frequency of fallow increases, the amount of soil organic matter decreases.

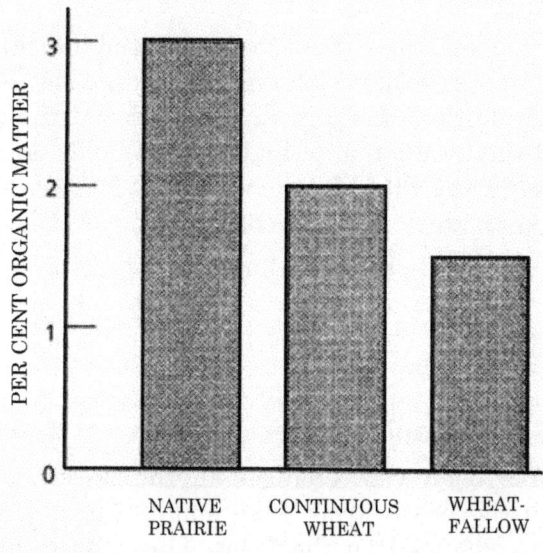

Figure: effect of frequency of fallow on per cent organic matter.

Summerfallowing for moisture conservation may be a necessary practice in the Brown and Dark Brown soil zones. However, it must be questioned in the Black and the Gray soil zones. Periodic fallowing may be acceptable in the higher rainfall regions for control of persistent perennial weeds and volunteer grains in pedigreed seed production. In the crop-fallow system common to the prairie region, the nitrogen removed has far exceeded that gained from crop residues, manure, legumes and fertilizer. The large reserves of nitrogen present in the organic matter of our prairie soils have been the major source of nitrogen in this cropping system. Continued reliance on soil organic matter reserves to supply the nitrogen requirements of crops will ultimately lead to a decline in soil productivity, and increased soil erosion.

When a change from a cropping system involving fallow to continuous cereal grain production, the nitrogen requirement increases. The nitrogen requirement is greatest in the first few years of continuous cropping as the nutrient cycling process adjusts to the new cropping system.

Crop Rotations

The value of forage crops in rotations with cereals and oilseeds has long been recognised, especially in the Luvisolic soils. Several long-term crop rotation studies conducted in Western Canada have shown that crop rotations involving perennial forages tend to stabilise soil organic matter at a higher level than crop rotations involving summerfallow.

The Breton Plots compared a two-year fallow-wheat rotation with a five-year rotation involving wheat, oats and barley followed by two years of hay production.

In research done by the University of Manitoba, the effects of various cultural practices on the level of organic matter are compared. It is interesting to note that the highest level of soil organic matter was maintained under continuous cropping.

The beneficial effects of perennial forages are the result of:
- a more extensive root system and crop aftermath contributing more organic matter to the soil,
- the fibrous nature of the root system of perennial grasses. These are particularly effective as a binding agent in soil aggregation,
- Nitrogen fertility enhancement by the growth of legumes,
- increased permeability of dense subsoils because of the deep penetrating tap roots of perennial legumes, especially alfalfa.
- a reduced rate of organic matter decomposition in the absence of tillage.

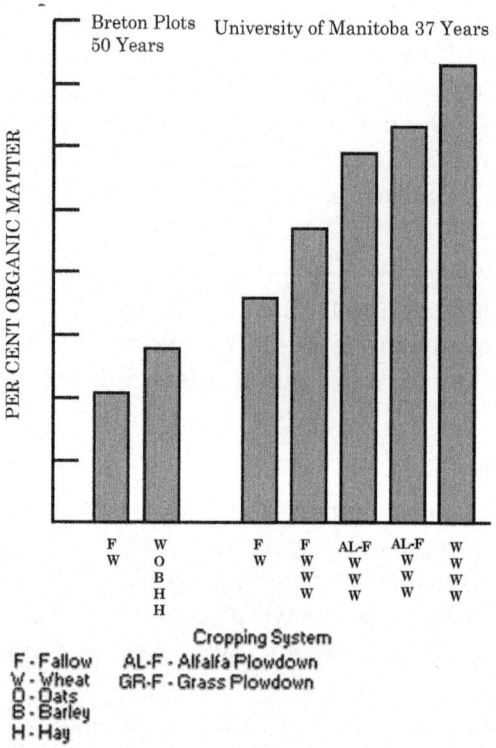

Figure: Effect of rotation on soil organic matter

In the brown and dark brown soil zones, perennial forages grown for forage production or as a plowdown crop may jeopardize subsequent cereal crops because they deplete soil moisture reserves.

Fertilization

Fertilizers will generally increase soil organic matter because the increased crop growth returns larger amounts of residues to the soil.

Data obtained from the Breton Plots. The increase in organic matter is less than what might be expected with current farming practices since all the straw had been removed from the plots.

To determine the fertilizer effect, two fertilizer treatments were averaged, one involving a low rate of nitrogen and sulphur and another involving a low rate of nitrogen, phosphorus, potassium and sulphur.

One would expect that with higher rates of fertilization, higher yielding varieties, and the return of all crop residues, the effect of fertilizers on organic matter would be greater.

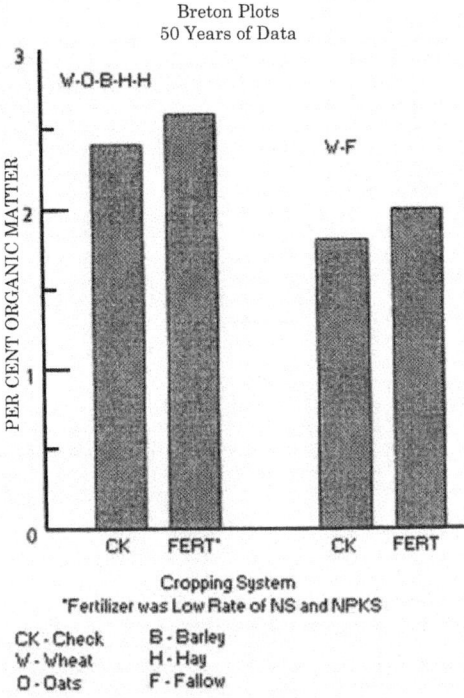

Figure: Effect of fertilizer on soil organic matter.

Plowdown

Legume plowdown has received considerable attention in recent years as an alternative to the use of nitrogen fertilizers. However, when considering this option in a cropping program, the amount of nitrogen added by the legume, as well as the loss of one year of production, the cost of seed and the expected yield increase must be kept in mind. Strictly as a source of nitrogen, the value of a legume plowdown is questionable. The amount of nitrogen fixed by a legume is dependent upon the type of legume, the amount of vegetative growth, the nature of the soil and environmental conditions. As a source of organic matter, legume plowdown is valuable, however, perennial forage is more effective than legume plowdown for increasing soil organic matter. Nitrate nitrogen which accumulates following legume plowdown is subject to loss, particularly in wet, poorly drained soils. To minimise this, legumes should be plowed down in the fall rather than mid-summer to reduce nitrate accumulation and subsequent loss.

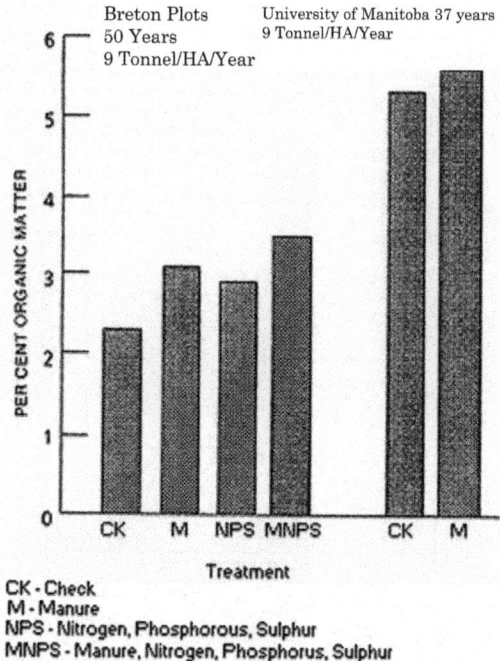

Figure: Effect of manure on soild organic matter.

The cultivation of prairie soils has generally resulted in a decline in organic matter of 30 to 50 per cent. A product of this decline has been the release of large amounts of plant nutrients, particularly nitrogen. Crop rotations with a high frequency of summerfallow have relied on the nitrogen released from soil organic matter to supply crop requirements.

More frequent or continuous cropping, less frequent tillage, the production of high yields and the return of crop residues will help to maintain soil organic matter at a satisfactory level. Perennial forages are effective for maintaining or increasing soil organic matter.

Biomass

Biomass, as a renewable energy source, is biological material from living, or recently living organisms. As an energy source, biomass can either be used directly, or converted into other energy products such as biofuel.

In the first sense, biomass is plant matter used to generate electricity with steam turbines & gasifiers or produce heat, usually by direct combustion. Examples include forest residues (such as dead trees, branches and tree stumps), yard clippings, wood chips and even municipal solid waste. In the second sense, biomass includes plant or animal matter that can be converted into fibres or other industrial chemicals, including biofuels. Industrial biomass can be grown from numerous types of plants, including miscanthus, switchgrass, hemp, corn, poplar, willow, sorghum, sugarcane, and a variety of tree species, ranging from eucalyptus to oil palm (palm oil).

Biomass Sources

Biomass is carbon, hydrogen and oxygen based. Biomass energy is derived from five distinct energy sources: garbage, wood, waste, landfill gases, and alcohol fuels. Wood energy is derived both from direct use of harvested wood as a fuel and from wood waste streams. The largest source of energy from wood is pulping liquor or "black liquor," a waste product from processes of the pulp, paper and paperboard industry. Waste energy is the second-largest source of biomass energy. The main contributors of waste energy are municipal solid waste (MSW), manufacturing waste, and landfill gas. Biomass alcohol fuel, or ethanol, is derived primarily from sugarcane and corn. It can be used directly as a fuel or as an additive to gasoline.

Biomass can be converted to other usable forms of energy like methane gas or transportation fuels like ethanol and biodiesel. Rotting garbage, and agricultural and human waste, release methane gas - also called "landfill gas" or "biogas." Crops like corn and sugar cane can be fermented to produce the transportation fuel, ethanol. Biodiesel, another transportation fuel, can be produced from left-over food products like vegetable oils and animal fats. Also, Biomass to liquids (BTLs) and cellulosic ethanol are still under research.

The biomass used for electricity production ranges by region. Forest by products, such as wood residues, are popular in the United States. Agricultural waste is common in Mauritius (sugar cane residue) and Southeast Asia (rice husks). Animal husbandry residues, such as poultry litter, is popular in the UK.

Biomass Conversion Process to Useful Energy

There are a number of technological options available to make use of a wide variety of biomass types as a renewable energy source. Conversion technologies may release the energy directly, in the form of heat or electricity, or may convert it to another form, such as liquid biofuel or combustible biogas. While for some classes of biomass resource there may be a number of usage options, for others there may be only one appropriate technology.

Thermal Conversion

These are processes in which heat is the dominant mechanism to convert the biomass into another chemical form. The basic alternatives of combustion, torrefaction, pyrolysis, and gasification are separated principally by the extent to which the chemical reactions involved are allowed to proceed (mainly controlled by the availability of oxygen and conversion temperature).

There are a number of other less common, more experimental or proprietary thermal processes that may offer benefits such as hydrothermal upgrading (HTU) and hydroprocessing.

Some have been developed for use on high moisture content biomass, including aqueous slurries, and allow them to be converted into more convenient forms.

Some of the applications of thermal conversion are combined heat and power (CHP) and co-firing. In a typical biomass power plant, efficiencies range from 20–27%.

Chemical Conversion

A range of chemical processes may be used to convert biomass into other forms, such as to produce a fuel that is more conveniently used, transported or stored, or to exploit some property of the process itself.

Biochemical Conversion

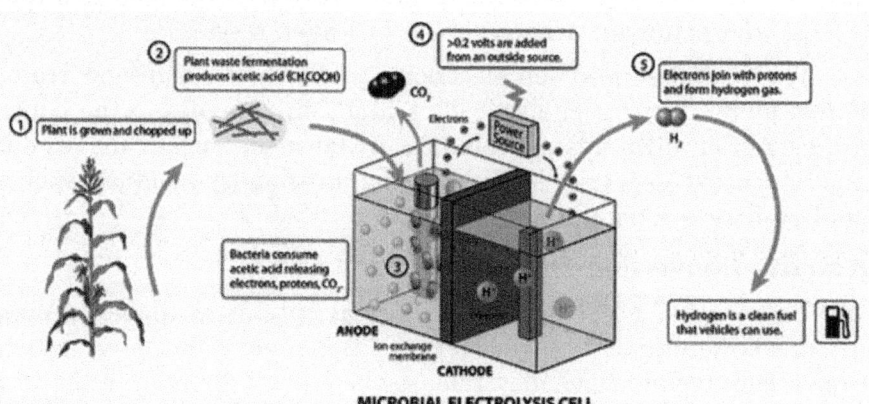

Figure: A microbial electrolysis cell can be used to directly make hydrogen gas from plant matter

As biomass is a natural material, many highly efficient biochemical processes have developed in nature to break down the molecules of which biomass is composed, and many of these biochemical conversion processes can be harnessed.

Biochemical conversion makes use of the enzymes of bacteria and other micro-organisms to break down biomass. In most cases micro-organisms are used to perform the conversion process: an aerobic digestion, fermentation and composting. Other chemical processes such as converting straight and waste vegetable oils into biodiesel is transesterification. Another way of breaking down biomass is by breaking down the carbohydrates and simple sugars to make alcohol. However, this process has not been perfected yet. Scientists are still researching the effects of converting biomass.

Environmental Impact

The existing biomass power generating industry in the United States, which consists of approximately 11,000 MW of summer operating capacity actively supplying power to the grid, produces about 1.4 percent of the U.S. electricity supply.

Currently, the New Hope Power Partnership is the largest biomass power plant in North America. The 140 MW facility uses sugar cane fibre (bagasse) and recycled urban wood as fuel to generate enough power for its large milling and refining operations as well as to supply renewable electricity for nearly 60,000 homes. The facility reduces dependence on oil by more than one million barrels per year, and by recycling sugar cane and wood waste, preserves landfill space in urban communities in Florida.

Using biomass as a fuel produces air pollution in the form of carbon monoxide, NOx (nitrogen oxides), VOCs (volatile organic compounds), particulates and other pollutants, in some cases at levels above those from traditional fuel sources such as coal or natural gas. Black carbon - a pollutant created by incomplete combustion of fossil fuels, biofuels, and biomass - is possibly the second largest contributor to global warming.

In 2009 a Swedish study of the giant brown haze that periodically covers large areas in South Asia determined that it had been principally produced by biomass burning, and to a lesser extent by fossil-fuel burning. Researchers measured a significant concentration of C, which is associated with recent plant life rather than with fossil fuels.

Biomass power plant size is often driven by biomass availability in close proximity as transport costs of the (bulky) fuel play a key factor in the plant's economics. It has to be noted, however, that rail and especially shipping on waterways can reduce transport costs significantly, which has led to a global biomass market. To make small plants of 1 MW_{el} economically profitable those power plants have need to be equipped with technology that is able to convert biomass to useful electricity with high efficiency such as ORC technology, a cycle similar to the water steam power process just with an organic working medium. Such small power plants can be found in Europe.

On combustion, the carbon from biomass is released into the atmosphere as carbon dioxide (CO_2). The amount of carbon stored in dry wood is approximately 50% by weight. When from agricultural sources, plant matter used as a fuel can be replaced by planting for new growth. When the biomass is from forests, the time to recapture the carbon stored is generally longer, and the carbon storage capacity of the forest may be reduced overall if destructive forestry techniques are employed.

Despite harvesting, biomass crops may sequester carbon. So for example soil organic carbon has been observed to be greater in switchgrass stands than in cultivated cropland soil, especially at depths below 12 inches. The grass sequesters the carbon in its increased root biomass. Typically, perennial crops sequester much more carbon than annual crops due to much greater non-harvested living biomass, both living and dead, built up over years, and much less soil disruption in cultivation.

The biomass-is-carbon-neutral proposal put forward in the early 1990s has been superseded by more recent science that recognises that mature, intact forests sequester carbon more effectively than cut-over areas. When a tree's carbon is released into the atmosphere in a single pulse, it contributes to climate change much more than woodland timber rotting slowly over decades. Current studies indicate that "even after 50 years the forest has not recovered to its initial carbon storage" and "the optimal strategy is likely to be protection of the standing forest".

Detritus

In biology, detritus is non-living particulate organic material (as opposed to dissolved organic material). It typically includes the bodies or fragments of dead organisms as well as fecal material. Detritus is typically colonised by communities of microorganisms which act to decompose (or remineralise) the material. In terrestrial ecosystems, it is encountered as leaf litter and other organic matter intermixed with soil, which is referred to as humus. Detritus of aquatic ecosystems is organic material suspended in water, which is referred to as marine snow.

Theory

Dead plants or animals, material derived from animal tissues (such as skin cast off during moulting and excreta) gradually lose their form, due to both physical processes and the action of decomposers, including grazers, bacteria and fungi. Decomposition, the process through which organic matter is decomposed, takes place in many stages. Materials like proteins, lipids and sugars with low molecular weight are rapidly consumed and absorbed by micro-organisms and organisms that feed on dead matter. Other compounds, such as complex carbohydrates are broken down more slowly. The various micro-organisms involved in the decomposition break down the organic materials in order to gain the resources they require for their own survival and proliferation. Accordingly, at the same time that the materials of plants and animals are being broken down, the materials (biomass) making up the bodies of the micro-organisms are built up by a process of assimilation. When micro-organisms die, fine organic particles are produced, and if these are eaten by small animals which feed on micro-organisms, they will collect inside the intestine, and change shape into large pellets of dung. As a result of this process, most of the materials from dead organisms disappears from view and is not obviously present in any recognisable form, but is in fact present in the form of a combination of fine organic particles and the organisms using them as nutrients. This combination is detritus.

In ecosystems on land, detritus is deposited on the surface of the ground, taking forms such as the humic soil beneath a layer of fallen leaves. In aquatic ecosystems, most detritus is suspended in water, and gradually settles. In particular, many different types of material are collected together by currents, and much material settles in slowly-flowing areas.

Much detritus is used as a source of nutrition for animals. In particular, many bottom feeding animals (benthos) living in mud flats feed in this way. In particular, since excreta are materials which other animals do not need, whatever energy value they might have, they are often unbalanced as a source of nutrients, and are not suitable as a source of nutrition on their own. However, there are many micro-organisms which multiply in natural environments.

These micro-organisms do not simply absorb nutrients from these particles, but also shape their own bodies so that they can take the resources they lack from the area around them, and this allows them to make use of excreta as a source of nutrients. In practical terms, the most important constituents of detritus are complex carbohydrates, which are persistent (difficult to break down), and the micro-organisms which multiply using these absorb carbon from the detritus, and materials such as nitrogen and phosphorus from the water in their environment to synthesize the components of their own cells.

A characteristic type of food chain called the detritus cycle takes place involving detritus feeders (detritivores), detritus and the micro-organisms that multiply on it.

For example, mud flats are inhabited by many univalves which are detritus feeders, such as moon shells. When these detritus feeders take in detritus with micro-organisms multiplying on it, they mainly break down and absorb the micro-organisms, which are rich in proteins, and excrete the detritus, which is mostly complex carbohydrates, having hardly broken it down at all.

At first this dung is a poor source of nutrition, and so univalves pay no attention to it, but after several days, micro-organisms begin to multiply on it again, its nutritional balance improves, and so they eat it again. Through this process of eating the detritus many times over and harvesting the micro-organisms from it, the detritus thins out, becomes fractured and becomes easier for the micro-organisms to use, and so the complex carbohydrates are also steadily broken down and disappear over time.

What is left behind by the detritivores is then further broken down and recycled by decomposers, such as bacteria and fungi.

This detritus cycle plays a large part in the so-called purification process, whereby organic materials carried in by rivers is broken down and disappears, and an extremely important part in the breeding and growth of marine resources. In ecosystems on land, far more essential material is broken down as dead material passing through the detritus chain than is broken down by being eaten by animals in a living state. In both land and aquatic ecosystems, the role played by detritus is too large to ignore.

Aquatic Ecosystems

In contrast to land ecosystems, dead materials and excreta in aquatic ecosystems do not settle immediately, and the finer the particles involved are, the longer they tend to take.

Terrestrial Ecosystems

Detritus occurs in a variety of terrestrial habitats including forest, chaparral and grassland. In forests the detritus is typically dominated by leaf, twig, and bacteria litter as measured by biomass dominance. There the leaf litter provides important cover for seedling protection as well as cover for a variety of arthropods, reptiles and amphibians. Some insect larvae feed on the detritus. Fungi and bacteria continue the decomposition process after grazers have consumed larger elements of the organic materials, and animal trampling has assisted in mechanically breaking down organic matter. At the later stages of decomposition, mesophilic micro-organisms decompose residual detritus, generating heat from exothermic processes; such heat generation is associated with the well known phenomenon of the elevated temperature of composting.

Consumers

There are an extremely large number of detritus feeders in water. After all, a large quantity of material is carried in by water currents. Even if an organism stays in a fixed position, as long as it has a system for filtering water, it will be able to obtain enough food to get by. Many rooted organisms survive in this way, using developed gills or tentacles to filter the water to take in food, a process known as filter feeding.

Another more widely used method of feeding, which also incorporates filter feeding, is a system where an organism secretes mucus to catch the detritus in lumps, and then carries these to its mouth using an area of cilia. This is called mucus feeding.

Many organisms, including sea slugs and serpent's starfish, scoop up the detritus which has settled on the water bed. Bivalves which live inside the water bed do not simply suck in water through their tubes, but also extend them to fish for detritus on the surface of the bed.

Producers

In contrast, from the point of view of organisms using photosynthesis, such as plants and plankton, detritus reduces the transparency of the water and gets in the way of their photosynthesis. However, given that they also require a supply of nutrient salts, in other words fertilizer for photosynthesis, their relationship with detritus is a complex one.

In land ecosystems, the waste products of plants and animals collect mainly on the ground (or on the surfaces of trees), and as decomposition proceeds, plants are supplied with fertilizer in the form of inorganic salts. However, in water, relatively little waste collects on the water bed, and so the progress of decomposition in water takes a more important role. However, investigating the level of inorganic salts in sea ecosystems shows that, unless there is an especially large supply, the quantity increases from winter to spring but is normally extremely low in summer. In line with this, the quantity of seaweed present reaches a peak in early summer, and then decreases. This is thought to be because organisms like plants grow quickly in warm periods and the quantity of inorganic salts is not enough to keep up with the demand. In other words, during winter, plant-like organisms are inactive and collect fertilizer, but if the temperature rises to some extent, they use this up in a very short period.

However, it is not the case that their productivity falls during the warmest periods. Organisms such as dinoflagellate have mobility, the ability to take in solid food, and the ability to photosynthesize. This type of micro-organism can take in substances such as detritus to grow, without waiting for it to be broken down into fertilizer.

Aquariums

In recent years, the word detritus has also come to be used in relation to aquariums (the word "aquarium" is a general term for any installation for keeping aquatic animals).

When animals such as fish are kept in an aquarium, substances such as excreta, mucus and dead skin cast off during moulting are produced by the animals and, naturally, generate detritus, and are continually broken down by micro-organisms.

Modern sealife aquariums often use the Berlin Method, which employs a piece of equipment called a protein skimmer, which produces air bubbles which the detritus adheres to, and forces it outside the tank before it decomposes, and also a highly porous type of natural rock called live rock where many bentos and bacteria live (hermatype which has been dead for some time is often used), which causes the detritus-feeding bentos and micro-organisms to undergo a detritus cycle. The Monaco system, where an an aerobic layer is created in the tank, to denitrify the organic compounds in the tank, and also the other nitrogen compounds, so that the decomposition process continues until the stage where water, carbon dioxide and nitrogen are produced, has also been implemented.

Initially, the filtration systems in water tanks often worked as the name suggests, using a physical filter to remove foreign substances in the water. Following this, the standard method for maintaining the water quality was to convert ammonium or nitrates in excreta, which have a high degree of neurotoxicity, but the combination of detritus feeders, detritus and micro-organisms has now brought aquarium technology to a still higher level.

Chapter 6
Organic Geochemistry

Organic geochemistry is the study of the impacts and processes that organisms have had on the Earth. The study of organic geochemistry is usually traced to the work of Alfred E. Treibs, "the father of organic geochemistry." Treibs first isolated metalloporphyrins from petroleum. This discovery established the biological origin of petroleum, which was previously poorly understood. Metalloorphyrins in general are highly stable organic compounds, and the detailed structures of the extracted derivatives made clear that they originated from chlorophyll.

The relationship between the occurrence of organic compounds in sedimentary deposits and petroleum deposits has long been of interest. Studies of ancient sediments and rock provide insights into the origins and sources of oil petroleum geochemistry and the biochemical antecedents of life.

Modern organic geochemistry includes studies of recent sediments to understand the carbon cycle, climate change, and ocean processes.

Total Organic Carbon

Total organic carbon (TOC) is the amount of carbon bound in an organic compound and is often used as a non-specific indicator of water quality or cleanliness of pharmaceutical manufacturing equipment.

A typical analysis for TOC measures both the total carbon present as well as the so called "inorganic carbon" (IC), the latter representing the content of dissolved carbon dioxide and carbonic acid salts. Subtracting the inorganic carbon from the total carbon yields TOC. Another common variant of TOC analysis involves removing the IC portion first and then measuring the leftover carbon. This method involves purging an acidified sample with carbon-free air or nitrogen prior to measurement, and so is more accurately called non-purgeable organic carbon (NPOC).

TOC Analysis

Environmental

Since the early 1970s, TOC has been recognised as an analytic technique to measure water quality during the drinking water purification process. TOC in source waters comes from decaying natural organic matter (NOM) and from synthetic sources. Humic acid, fulvic acid, amines, and urea are types of NOM. Detergents, pesticides, fertilizers, herbicides, industrial chemicals, and chlorinated organics are examples of synthetic sources. Before source water is treated for disinfection, TOC provides an important role in quantifying the amount of NOM in the water source. In water treatment facilities, source water is subject to reaction with chloride containing disinfectants. When the raw water is chlorinated, active chlorine compounds (Cl_2, HOCl, ClO) react with NOM to produce chlorinated disinfection byproducts (DBPs). Many researchers have determined that higher levels of NOM in source water during the disinfection process will increase the amount of carcinogens in the processed drinking water.

With passage of the U.S. Safe Drinking Water Act in 1974, TOC analysis emerged as a rapid and accurate alternative to the classical but lengthy biological oxygen demand (BOD) and chemical oxygen demand (COD) tests traditionally reserved for assessing the pollution potential of wastewaters. Today, environmental agencies regulate the trace limits of DBPs in drinking water. Recently published analytical methods, such as United States Environmental Protection Agency (EPA) method 415.3, support the Agency's *Disinfectants and Disinfection Byproducts Rules,* which regulate the amount of NOM to prevent the formation of DBPs in finished waters.

Pharmaceutical

Introduction of organic matter into water systems occurs not only from living organisms and from decaying matter in source water, but also from purification and distribution system materials. A relationship may exist between endotoxins, microbial growth, and the development of biofilms on pipeline walls and biofilm growth within pharmaceutical distribution systems. A correlation is believed to exist between TOC concentrations and the levels of endotoxins and microbes. Sustaining low TOC levels helps to control levels of endotoxins and microbes and thereby the development of biofilm growth. The United States Pharmacopoeia (USP), European Pharmacopoeia (EP) and Japanese Pharmacopoeia (JP) recognise TOC as a required test for purified water and water for injection (WFI).

For this reason, TOC has found acceptance as a process control attribute in the biotechnology industry to monitor the performance of unit operations comprising purification and distribution systems. As many of these biotechnology operations include the preparation of medicines, the U.S. Food and Drug Administration (FDA) enacts numerous regulations to protect the health of the public and ensure the product quality is maintained. To make sure there is no cross-contamination between product runs of different drugs, various cleaning procedures are performed. TOC concentration levels are used to track the success of these cleaning validation procedures especially clean-in-place (CIP).

Measurement

To understand the analysis process better, some key basic terminologies should be understood and their relationships to one another.

- Total Carbon (TC) – all the carbon in the sample, including both inorganic and organic carbon
- Total Inorganic Carbon (TIC) – often referred to as inorganic carbon (IC), carbonate, bicarbonate, and dissolved carbon dioxide (CO_2).
- Total Organic Carbon (TOC) – material derived from decaying vegetation, bacterial growth, and metabolic activities of living organisms or chemicals.
- Non-Purgeable Organic Carbon (NPOC) – commonly referred to as TOC; organic carbon remaining in an acidified sample after purging the sample with gas.
- Purgeable (volatile) Organic Carbon (VOC) – organic carbon that has been removed from a neutral, or acidified sample by purging with an inert gas. These are the same compounds referred to as Volatile Organic Compounds (VOC) and usually determined by Purge and Trap Gas Chromatography.
- Dissolved Organic Carbon (DOC) – organic carbon remaining in a sample after filtering the sample, typically using a 0.45 micrometre filter.

- Suspended Organic Carbon – also called particulate organic carbon (POC); the carbon in particulate form that is too large to pass through a filter.

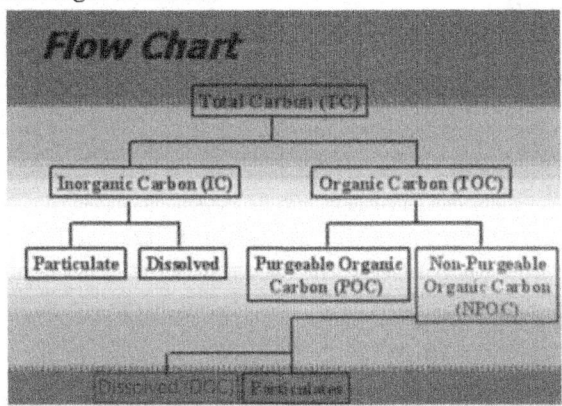

Figure: Flow Chart displaying the relationship of carbon components

Since all TOC analysers only actually measure total carbon, TOC analysis always requires some accounting for the inorganic carbon that is always present. One analysis technique involves a two-stage process commonly referred to as TC-IC. It measures the amount of inorganic carbon (IC) evolved from an acidified aliquot of a sample and also the amount of total carbon (TC) present in the sample. TOC is calculated by subtraction of the IC value from the TC the sample. Another variant employs acidification of the sample to evolve carbon dioxide and measuring it as inorganic carbon (IC), then oxidising and measuring the remaining non-purgeable organic carbon (NPOC). This is called TIC-NPOC analysis. A more common method directly measures TOC in the sample by again acidifying the sample it to a pH value of two or less to release the IC gas but in this case to the air not for measurement. The remaining non-purgeable CO_2 gas (NPOC) contained in the liquid aliquot is then oxidised releasing the gases. These gases are then sent to the detector for measurement.

Whether the analysis of TOC is by TC-IC or NPOC methods, it may be broken into three main stages:

1. Acidification
2. Oxidation
3. Detection and Quantification.

The first stage is acidification of the sample for the removal of the IC and POC gases. The release of these gases to the detector for measurement or to the air is dependent upon which type of analysis is of interest, the former for TC-IC and the latter for TOC (NPOC).

Acidification

The removal and venting of IC and POC gases from the liquid sample by acidification and sparging occurs in the following manner.

$$HCO_3^-, CO_3^{-2}, CO_2, POC \xrightarrow[\text{carrier gas, acid}]{} CO_2 + POC$$

Oxidation

The second stage is the oxidation of the carbon in the remaining sample in the form of carbon dioxide (CO_2) and other gases. Modern TOC analysers perform this oxidation step by several processes:

1. High Temperature Combustion
2. High temperature catalytic (HTCO) oxidation
3. Photo-oxidation alone
4. Thermo-chemical oxidation
5. Photo-chemical oxidation
6. Electrolytic Oxidation.

High Temperature Combustion

Prepared samples are combusted at 1,350 °C in an oxygen-rich atmosphere. All carbon present converts to carbon dioxide, flows through scrubber tubes to remove interferences such as chlorine gas, and water vapour, and the carbon dioxide is measured either by absorption into a strong base then weighed, or using an Infrared Detector. Most modern analysers use non-dispersive infrared (NDIR) for detection of the carbon dioxide.

High Temperature Catalytic Oxidation

A manual or automated process injects the sample onto a platinum catalyst at 680 °C in an oxygen rich atmosphere. The concentration of carbon dioxide generated is measured with a non-dispersive infrared (NDIR) detector.

Oxidation of the sample is complete after injection into the furnace, turning oxidisable material in the sample into gaseous form. A carbon-free carrier gas transports the CO_2, through a moisture trap and halide scrubbers to remove water vapour and halides from the gas stream before it reaches the detector. These substances can interfere with the detection of the CO_2 gas. The HTCO method may be useful in those applications where difficult to oxidise compounds, or high molecular weight organics, are present as it provides almost complete oxidation of organics including solids and particulates small enough to be injected into the furnace.

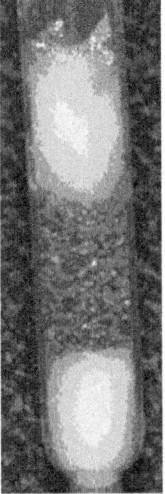

Figure: A HTCO combustion tube packed with platinum catalyst

The major drawback of HTCO analysis is its unstable baseline resulting from the gradual accumulation of non-volatile residues within the combustion tube. These residues continuously change TOC background levels requiring continuous background correction. Because aqueous samples are injected directly into a very hot, usually quartz, furnace only small aliquots (less than 2 millilitres and usually less than 400 micro-litres) of sample can be handled making the methods less sensitive than chemical oxidation methods capable of digesting as much as 10 times more sample.

Also, the salt content of the samples do not combust, and so therefore, gradually build a residue inside the combustion tube eventually clogging the catalyst resulting in poor peak shapes, and degraded accuracy or precision, unless appropriate maintenance procedures are followed. The catalyst should be regenerated or replaced as needed.

Photo-Oxidation (UV Light)

In this oxidation scheme, ultraviolet light alone oxidises the carbon within the sample to produce CO_2. The UV oxidation method offers the most reliable, low maintenance method of analysing TOC in ultra-pure waters.

The UV/Chemical (Persulfate) Oxidation

Like the photo-oxidation method, UV light is the oxidiser but the oxidation power of the reaction is magnified by the addition of a chemical oxidiser, which is usually a persulfate compound. The mechanisms of the reactions are as follows:

Free radical Oxidants formed:

$$S_2O_8^{2-} \xrightarrow{hv} 2SO_4^{-\cdot}$$

$$H_2O \xrightarrow{hv} H^+ + OH^{\cdot}$$

$$SO_4^{-\cdot} + H_2O \rightarrow SO_4^{2-} + OH^{\cdot} + H^+$$

Excitation of organics:

$$R \xrightarrow{hv} R^*$$

Oxidation of organics:

$$R^* + SO_4^{-\cdot} + OH^{\cdot} \rightarrow nCO_2 + ...$$

The UV–chemical oxidation method offers a relatively low maintenance, high sensitivity method for a wide range of applications. However, there are oxidation limitations of this method. Limitations include the inaccuracies associated with the addition of any foreign substance into the analyte and samples with high amounts of particulates. Performing "System Blank" analysis, which is to analyse then subtract the amount of carbon contributed by the chemical additive, inaccuracies are lowered. However, analyses of levels below 200 ppb TOC are still difficult.

Thermo-Chemical (Persulfate) Oxidation

Also known as heated persulfate, the method utilises the same free radical formation as UV persulfate oxidation except uses heat to magnify the oxidising power of persulfate. Chemical oxidation of carbon with a strong oxidiser, such as persulfate, is highly efficient, and unlike UV, is not susceptible to lower recoveries caused by turbidity in samples. The analysis of system blanks, necessary in all chemical procedures, is especially necessary with heated persulfate TOC methods because the method is so sensitive that reagents cannot be prepared with carbon contents low enough to not be detected. Persulfate methods are used in the analysis of wastewater, drinking water, and pharmaceutical waters. When used in conjunction with sensitive NDIR detectors heated persulfate TOC instruments readily measure TOC at single digit parts per billion (ppb) up to hundreds of parts per million (ppm) depending on sample volumes.

Detection and Quantification

Accurate detection and quantification are the most vital components of the TOC analysis process. Conductivity and non-dispersive infrared (NDIR) are the two common detection methods used in modern TOC analysers.

Conductivity

There are two types of conductivity detectors, direct and membrane. Direct conductivity provides an inexpensive and simple means of measuring CO_2. This method has good oxidation of organics, uses no carrier gas, is good at the parts per billion (ppb) ranges, but has a very limited analytical range. Membrane conductivity relies upon the same technology as direct conductivity. Although it is more robust than direct conductivity, it suffers from slow analysis time. Both methods analyse sample conductivity before and after oxidisation, attributing this differential measurement to the TOC of the sample. During the sample oxidisation phase, CO_2 (directly related to the TOC in the sample) and other gases are formed.

The dissolved CO_2 forms a weak acid, thereby changing the conductivity of the original sample proportionately to the TOC in the sample. Conductivity analyses assume that only CO_2 is present within the solution. As long as this holds true, then the TOC calculation by this differential measurement is valid. However, depending on the chemical species present in the sample and their individual products of oxidation, they may present either a positive or a negative interference to the actual TOC value, resulting in analytical error. Some of the interfering chemical species include Cl^-, HCO_3^-, SO_3^{2-}, SO_2^-, ClO_2^-, and H^+.

Small changes in pH and temperature fluctuations also contribute to inaccuracy. Membrane conductivity analysers have tried to improve upon the direct conductivity approach by incorporating the use of hydrophobic gas permeation membranes to allow a more "selective" passage of the dissolved CO_2 gas. While this has solved certain problems, membranes have their own particular limitations, such as with true selectivity, clogging and, more undetectably, they provide secondary sites for other chemical reactions, which are prone to display "false negatives," a condition far more severe than "false positives" in critical applications. Micro leaks, flow problems, dead spots, microbial growth (blockage) are also potential problems. Most disconcerting is the inability of membrane methods to recover to operational performance after an overload or "spill" condition arises to over range the instrument, often taking hours before returning to reliable service and recalibration, just when accuracy of TOC analysis is most critical to operators for quality control.

Non-dispersive Infrared (NDIR)

The non-dispersive infrared analysis (NDIR) method offers the only practical interference-free method for detecting CO_2 in TOC analysis. The principal advantage of using NDIR is that it directly and specifically measures the CO_2 generated by oxidation of the organic carbon in the oxidation reactor, rather than relying on a measurement of a secondary, corrected effect, such as used in conductivity measurements. A traditional NDIR detector relies upon flow-through-cell technology, the oxidation product flows into and out of the detector continuously. A region of adsorption of infrared light specific to CO_2, usually around 4.26 μm (2350 cm^{-1}), is measured over time as the gas flows through the detector. The infrared absorption spectra of CO_2 and other gases . A second reference measurement that is non-specific to CO_2 is also taken and the differential result correlates to the CO_2 concentration in the detector at that moment. As the gas continues to flow into and out of the detector cell the sum of the measurements results in a peak that is integrated and correlated to the total CO_2 concentration in the sample aliquot.

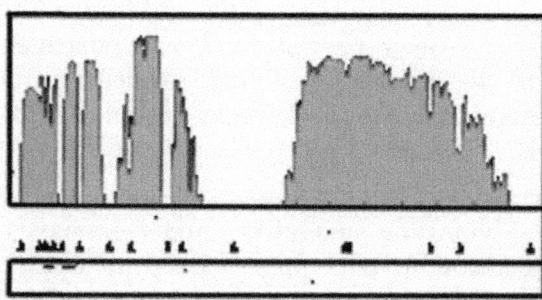

Figure: Plot of atmospheric transmittance in part of IR region showing CO_2 absorbing wavelengths

Recent Advances in NDIR Technology

A new advance of NDIR technology is Static Pressurised Concentration (SPC).

The exit valve of the NDIR is closed to allow the detector to become pressurised. Once the gases in the detector have reached equilibrium, the concentration of the CO_2 is analysed. This pressurisation of the sample gas stream in the NDIR, a patent-pending technique, allows for increased sensitivity and precision by measuring the entirety of the oxidation products of the sample in one reading, compared to flow-through cell technology. The output signal is proportional to the concentration of CO_2 in the carrier gas, from the oxidation of the sample aliquot. UV/ Persulfate oxidation combined with NDIR detection provides good oxidation of organics, low instrument maintenance, good precision at ppb levels, relatively fast sample analysis time and easily accommodates multiple applications, including purified water (PW), water for injection (WFI), CIP, drinking water and ultra-pure water analyses.

Analysers

Virtually all TOC analysers measure the CO_2 formed when organic carbon is oxidised and/or when inorganic carbon is acidified. Oxidation is performed either through Pt-catalysed combustion, by heated persulfate, or with a UV/persulfate reactor. Once the CO_2 is formed, it is measured by a detector: either a conductivity cell (if the CO_2 is aqueous) or a non-dispersive infrared cell (after purging the aqueous CO_2 into the gaseous phase). Conductivity detection is only desirable in the lower TOC ranges in deionized waters, whereas NDIR detection excels in all TOC ranges. A variation described as Membrane Conductivity Detection can allow for measurement of TOC across a wide analytical range in both deionized and non-deionized water samples. Modern high-performance TOC instruments are capable of detecting carbon concentrations well below 1 µg/L (1 part per billion or ppb).

A total organic carbon analyser determines the amount of carbon in a water sample. By acidifying the sample and flushing with nitrogen or helium the sample removes inorganic carbon, leaving only organic carbon sources for measurement. There are two types of analysers. One uses combustion and the other chemical oxidation. This is used as a water purity test, as the presence of bacteria introduces organic carbon.

Analyser Field Testing and Reports

A non-profit research and testing organisation, the Instrumentation Testing Association (ITA) offers a report that provides results of field testing online TOC analysers in an industrial wastewater application. Gulf Coast Waste Disposal Authority (GCWDA), Bayport Industrial Wastewater Treatment Plant in Pasadena, Texas sponsored and conducted this test in 2011. The GCWDA Bayport facility treats approximately 30 mgd of industrial waste received from approximately 65 customers (primarily petrochemical).

Field tests consisted of operating online TOC analysers at the influent of the Bayport facility in which TOC concentrations can range from 490 to 1020 mg/L with an average of 870 mg/L. GCWDA conducts approximately 102 TOC analyses in their laboratory per day at their Bayport treatment facility and use TOC measurements for process control and billing purposes. GCWDA plans to use online TOC analysers for process control, detecting influent slug loads from industries and to potentially use online TOC analysers to detect and monitor volatiles of the incoming stream.

Field tests were conducted for a period of 90-days and used laboratory conformance measurements once per day to compare with analyser output to demonstrate the instrument's overall accuracy when subjected to many simultaneously changing parameters as experienced in real-time monitoring conditions.

Field test results can provide information regarding instrument design, operation and maintenance requirements which influence the performance of the instruments in field applications. The field test report includes evaluations of online TOC analysers utilising the following technologies: High Temperature Combustion (HTC), High Temperature Catalytic/Combustion Oxidation (HTCO), Supercritical Water Oxidation (SCWO), and Two-Stage Advanced Oxidation (TSAO).

Combustion

In a combustion analyser, half the sample is injected into a chamber where it is acidified, usually with phosphoric acid, to turn all of the inorganic carbon into carbon dioxide as per the following reaction:

$$CO_2 + H_2O \rightarrow H_2CO_3 \rightarrow H^+ + HCO_3^- \rightarrow 2H^+ + CO_3^{2-}$$

This is then sent to a detector for measurement. The other half of the sample is injected into a combustion chamber which is raised to between 600–700°C, some even up to 1200°C. Here, all the carbon reacts with oxygen, forming carbon dioxide. It's then flushed into a cooling chamber, and finally into the detector. Usually, the detector used is a non-dispersive infrared spectrophotometre. By finding the total inorganic carbon and subtracting it from the total carbon content, the amount of organic carbon is determined.

Chemical Oxidation

Chemical oxidation analysers inject the sample into a chamber with phosphoric acid followed by persulfate. The analysis is separated into two steps. One removes inorganic carbon by acidification and purging. After removal of inorganic carbon persulfate is added and the sample is either heated or bombarded with UV light from a mercury vapour lamp.

Free radicals form from the persulfate and react with any carbon available to form carbon dioxide. The carbon from both determination (steps) is either run through membranes which measure the conductivity changes that result from the presence of varying amounts of carbon dioxide, or purged into and detected by a sensitive NDIR detector. Same as the combustion analyser, the total carbon formed minus the inorganic carbon gives a good estimate of the total organic carbon in the sample.

This method is often used in online applications because of its low maintenance requirements. For example the online Biotector which is the most modern application of this method.

Applications

TOC is the first chemical analysis to be carried out on potential petroleum source rock in oil exploration. It is very important in detecting contaminants in drinking water, cooling water, water used in semiconductor manufacturing, and water for pharmaceutical use. Analysis may be made either as an online continuous measurement or a lab-based measurement.

TOC detection is an important measurement because of the effects it may have on the environment, human health, and manufacturing processes. TOC is a highly sensitive, non-specific measurement of all organics present in a sample. It, therefore, can be used to regulate the organic chemical discharge to the environment in a manufacturing plant. In addition, low TOC can confirm the absence of potentially harmful organic chemicals in water used to manufacture pharmaceutical products. TOC is also of interest in the field of potable water purification due to disinfection of byproducts. Inorganic carbon poses little to no threat.

Walkely-Black (1934) Modified Method

Calculation

In the Total Organic Carbon determination to obtain the percent carbon content from soil, first we have to standardise the titrant solution ($FeSO_4 \cdot 7H_2O$) before the sample analysis are made, as result we obtain some data which have to be reduced in order to obtain the results we need, and to do this, we use the next equations:

Eq.1. Titrant normality equation:

$$N_2 = \frac{N_1 V_1}{V_2}$$

where:

N_1: $K_2Cr_2O_7$ normality

V_1: $K_2Cr_2O_7$ volume (mL)

V_2: $FeSO_4$ volume (mL)

Eq. 2. Organic carbon percentage:

$$\%Carbon = \frac{(A-B)*0.3*1.33}{C}$$

where:

A: meq $K_2Cr_2O_7$ = (mL $K_2Cr_2O_7$ x N $K_2Cr_2O_7$)

B: meq $FeSO_4 \cdot 7H_2O$ = (mL $FeSO_4 \cdot 7H_2O$ x N $FeSO_4 \cdot 7H_2O$)

C: grams of sample

0.3: Conversion factor to carbon weight.

We have milliequivalents as result of the difference between A and B, and they need to be converted to carbon milliequivalents in order to get the units we need, for that it is necessary to do the next operation:

Eq. 3

$meq * (1eq / 1000meq) * ((1/4)molC / 1eqC) * (12g / 1molC) * 100 = 0.3gC(\%)$

The 0.3 conversion factor has units of carbon grams and involves the constant to convert a fraction to percent units; hence equation 2 does not have the factor 100. Walkey-Black constant for sediments. 75% is the mean recuperation of carbon in solids and sediments by using this method, that's why the final result has to be multiplied by 1.33 in order to get the real value, this constant is not used when determining carbon in KHP standard because almost all its carbon content is recovered.

Biotic Material

Biotic material or biological derived material is any natural material that is originated from living organisms. Most such materials contain carbon and are capable of decay.

Examples of biotic materials are wood, linoleum, straw, humus, manure, bark, crude oil, cotton, spider silk, chitin, fibrin, and bone.

The use of biotic materials, and processed biotic materials (bio-based material) as alternative natural materials, over synthetics is popular with those who are environmentally conscious because such materials are usually biodegradable, renewable, and the processing is commonly understood and has minimal environmental impact. However, not all biotic materials are environmentally friendly, such as those that require high levels of processing, are harvested unsustainably, or are used to produce carbon emissions.

When the source of the recently-living material has little importance to the product produced, such as in the production of biofuels, biotic material is simply called biomass. Many fuel sources may have biological sources, and may be divided roughly into fossil fuels, and biofuel. In soil science, biotic material is often referred to as *organic matter*. Biotic materials in soil include glomalin, Dopplerite and humic acid. Some biotic material may not be considered to be organic matter if it is low in organic compounds, such as a clam's shell, which is an essential component of the living organism, but contains little organic carbon.

Examples of the use of biotic materials include:
- Alternative natural materials
- building material, for a stylistic reasons, or to reduce allergic reactions.
- clothing
- energy production
- food
- medicine
- ink
- composting and mulch.

Organic Farming

Organic farming is the form of agriculture that relies on techniques such as crop rotation, green manure, compost and biological pest control to maintain soil productivity and control pests on a farm. Organic farming uses fertilizers and pesticides but excludes or strictly limits the use of manufactured(synthetic) fertilizers, pesticides (which include herbicides, insecticides and fungicides), plant growth regulators such as hormones, livestock antibiotics, food additives, and genetically modified organisms.

Organic agricultural methods are internationally regulated and legally enforced by many nations, based in large part on the standards set by the International Federation of Organic Agriculture Movements (IFOAM), an international umbrella organisation for organic farming organisations established in 1972. IFOAM defines the overarching goal of organic farming as:

> *"Organic agriculture is a production system that sustains the health of soils, ecosystems and people. It relies on ecological processes, biodiversity and cycles adapted to local conditions, rather than the use of inputs with adverse effects. Organic agriculture combines tradition, innovation and science to benefit the shared environment and promote fair relationships and a good quality of life for all involved.."*

—International Federation of Organic Agriculture Movements

Since 1990, the market for organic products has grown from nothing, reaching $55 billion in 2009 according to Organic Monitor (www.organicmonitor.com). This demand has driven a similar increase in organically managed farmland. Approximately 37,000,000 hectares (91,000,000 acres) worldwide are now farmed organically, representing approximately 0.9 percent of total world farmland (2009).

History

Organic farming (of many particular kinds) was the original type of agriculture, and has been practiced for thousands of years. After the industrial revolution had introduced inorganic methods, some of which were not well developed and had serious side effects, an organic movement began in the 1940s as a reaction to agriculture's growing reliance on synthetic fertilizers. Artificial fertilizers had been created during the 18th century, initially with superphosphates and then ammonia-based fertilizers mass-produced using the Haber-Bosch process developed during World War I. These early fertilizers were cheap, powerful, and easy to transport in bulk. Similar advances occurred in chemical pesticides in the 1940s, leading to the decade being referred to as the 'pesticide era'.

Although organic farming is prehistoric in the widest sense, Sir Albert Howard is widely considered to be the "father of organic farming" in the sense that he was a key founder of the post-industrial-revolution organic movement.

Further work was done by J.I. Rodale in the United States, Lady Eve Balfour in the United Kingdom, and many others across the world. The modern organic movement is a revival movement in the sense that it seeks to restore balance that was lost when technology grew rapidly in the 19th and 20th centuries.

Modern organic farming has made up only a fraction of total agricultural output from its beginning until today. Increasing environmental awareness in the general population has transformed the originally supply-driven movement to a demand-driven one. Premium prices and some government subsidies attracted farmers. In the developing world, many producers farm according to traditional methods which are comparable to organic farming but are not certified. In other cases, farmers in the developing world have converted for economic reasons.

Methods

Figure: Organic cultivation of mixed vegetables in Capay, California. Note the hedgerow in the background.

"An organic farm, properly speaking, is not one that uses certain methods and substances and avoids others; it is a farm whose structure is formed in imitation of the structure of a natural system that has the integrity, the independence and the benign dependence of an organism"
—Wendell Berry, "The Gift of Good Land"

Soil Management

Plants need nitrogen, phosphorus, and potassium, as well as micronutrients and symbiotic relationships with fungi and other organisms to flourish, but getting enough nitrogen, and particularly synchronisation so that plants get enough nitrogen at the right time (when plants need it most), is likely the greatest challenge for organic farmers. Crop rotation and green manure ("cover crops") help to provide nitrogen through legumes (more precisely, the *Fabaceae* family) which fix nitrogen from the atmosphere through symbiosis with rhizobial bacteria. Intercropping, which is sometimes used for insect and disease control, can also increase soil nutrients, but the competition between the legume and the crop can be problematic and wider spacing between crop rows is required. Crop residues can be ploughed back into the soil, and different plants leave different amounts of nitrogen, potentially aiding synchronisation. Organic farmers also use animal manure, certain processed fertilizers such as seed meal and various mineral powders such as rock phosphate and greensand, a naturally occurring form of potash which provides potassium. Together these methods help to control erosion. In some cases pH may need to be amended. Natural pH amendments include lime and sulfur, but in the U.S. some compounds such as iron sulfate, aluminium sulfate, magnesium sulfate, and soluble boron products are allowed in organic farming.

Mixed farms with both livestock and crops can operate as ley farms, whereby the land gathers fertility through growing nitrogen-fixing forage grasses such as white clover or alfalfa and grows cash crops or cereals when fertility is established. Farms without livestock ("stockless") may find it more difficult to maintain fertility, and may rely more on external inputs such as imported manure as well as grain legumes and green manures, although grain legumes may fix limited nitrogen because they are harvested. Horticultural farms growing fruits and vegetables which operate in protected conditions are often even more reliant upon external inputs.

Biological research on soil and soil organisms has proven beneficial to organic farming. Varieties of bacteria and fungi break down chemicals, plant matter and animal waste into productive soil nutrients. In turn, they produce benefits of healthier yields and more productive soil for future crops. Fields with less or no manure display significantly lower yields, due to decreased soil microbe community, providing a healthier, more arable soil system.

Weed Management

Organic weed management promotes weed suppression, rather than weed elimination, by enhancing crop competition and phytotoxic effects on weeds. Organic farmers integrate cultural, biological, mechanical, physical and chemical tactics to manage weeds without synthetic herbicides.

Organic standards require rotation of annual crops, meaning that a single crop cannot be grown in the same location without a different, intervening crop. Organic crop rotations frequently include weed-suppressive cover crops and crops with dissimilar life cycles to discourage weeds associated with a particular crop. Organic farmers strive to increase soil organic matter content, which can support microorganisms that destroy common weed seeds.

Other cultural practices used to enhance crop competitiveness and reduce weed pressure include selection of competitive crop varieties, high-density planting, tight row spacing, and late planting into warm soil to encourage rapid crop germination.

Mechanical and physical weed control practices used on organic farms can be broadly grouped as:

- Tillage - Turning the soil between crops to incorporate crop residues and soil amendments; remove existing weed growth and prepare a seedbed for planting;
- Cultivation - Disturbing the soil after seeding;
- Mowing and cutting - Removing top growth of weeds;
- Flame weeding and thermal weeding - Using heat to kill weeds; and
- Mulching - Blocking weed emergence with organic materials, plastic films, or landscape fabric.

Some naturally sourced chemicals are allowed for herbicidal use. These include certain formulations of acetic acid (concentrated vinegar), corn gluten meal, and essential oils. A few selective bioherbicides based on fungal pathogens have also been developed. At this time, however, organic herbicides and bioherbicides play a minor role in the organic weed control toolbox.

Weeds can be controlled by grazing. For example, geese have been used successfully to weed a range of organic crops including cotton, strawberries, tobacco, and corn, reviving the practice of keeping cotton patch geese, common in the southern U.S. before the 1950s. Similarly, some rice farmers introduce ducks and fish to wet paddy fields to eat both weeds and insects.

Controlling other Organisms

Organisms aside from weeds that cause problems on organic farms include arthropods (e.g., insects, mites), nematodes, fungi and bacteria. Organic farmers use a wide range of Integrated Pest Management practices to prevent pests and diseases. These include, but are not limited to, crop rotation and nutrient management; sanitation to remove pest habitat; provision of habitat for beneficial organisms; selection of pest-resistant crops and animals; crop protection using physical barriers, such as row covers; and crop diversification through companion planting or establishment of polycultures.

Organic farmers often depend on biological pest control, the use of beneficial organisms to reduce pest populations. Examples of beneficial insects include minute pirate bugs, big-eyed bugs, and to a lesser extent ladybugs (which tend to fly away), all of which eat a wide range of pests. Lacewings are also effective, but tend to fly away. Praying mantis tend to move more slowly and eat less heavily. Parasitoid wasps tend to be effective for their selected prey, but like all small insects can be less effective outdoors because the wind controls their movement. Predatory mites are effective for controlling other mites.

When these practices are insufficient to prevent or control pests an organic farmer may apply a pesticide. With some exceptions, naturally occurring pesticides are allowed for use on organic farms, and synthetic substances are prohibited. Pesticides with different modes of action should be rotated to minimise development of pesticide resistance.

Naturally derived insecticides allowed for use on organic farms use include *Bacillus thuringiensis* (a bacterial toxin), pyrethrum (a chrysanthemum extract), spinosad (a bacterial metabolite), neem (a tree extract) and rotenone (a legume root extract). These are sometimes called green pesticides because they are generally, but not necessarily, safer and more environmentally friendly than synthetic pesticides. Rotenone and pyrethrum are particularly controversial because they work by attacking the nervous system, like most conventional insecticides. Fewer than 10% of organic farmers use these pesticides regularly; one survey found that only 5.3% of vegetable growers in California use rotenone while 1.7% use pyrethrum (Lotter 2003:26).

Naturally derived fungicides allowed for use on organic farms include the bacteria *Bacillus subtilis* and *Bacillus pumilus*; and the fungus *Trichoderma harzianum*. These are mainly effective for diseases affecting roots. Agricultural Research Service scientists have found that caprylic acid, a naturally occurring fatty acid in milk and coconuts, as well as other natural plant extracts have antimicrobial characteristics that can help. Compost tea contains a mix of beneficial microbes, which may attack or out-compete certain plant pathogens, but variability among formulations and preparation methods may contribute to inconsistent results or even dangerous growth of toxic microbes in compost teas.

Some naturally derived pesticides are not allowed for use on organic farms. These include nicotine sulfate, arsenic, and strychnine.

Synthetic pesticides allowed for use on organic farms include insecticidal soaps and horticultural oils for insect management; and Bordeaux mixture, copper hydroxide and sodium bicarbonate for managing fungi.

Genetic Modification

A key characteristic of organic farming is the rejection of genetically engineered plants and animals. On October 19, 1998, participants at IFOAM's 12th Scientific Conference issued the Mar del Plata Declaration, where more than 600 delegates from over 60 countries voted unanimously to exclude the use of genetically modified organisms in food production and agriculture.

Although opposition to the use of any transgenic technologies in organic farming is strong, agricultural researchers Luis Herrera-Estrella and Ariel Alvarez-Morales continue to advocate integration of transgenic technologies into organic farming as the optimal means to sustainable agriculture, particularly in the developing world.

Similarly, some organic farmers question the rationale behind the ban on the use of genetically engineered seed because they view this kind of biotechnology consistent with organic principles.

Although GMOs are excluded from organic farming, there is concern that the pollen from genetically modified crops is increasingly penetrating organic and heirloom seed stocks, making it difficult, if not impossible, to keep these genomes from entering the organic food supply. International trade restrictions limit the availability GMOs to certain countries.

The hazards that genetic modification could pose to the environment are hotly contested.

Standards

Standards regulate production methods and in some cases final output for organic agriculture. Standards may be voluntary or legislated. As early as the 1970s private associations certified organic producers. In the 1980s, governments began to produce organic production guidelines. In the 1990s, a trend toward legislated standards began, most notably with the 1991 EU-Eco-regulation developed for European Union, which set standards for 12 countries, and a 1993 UK program. The EU's program was followed by a Japanese program in 2001, and in 2002 the U.S. created the National Organic Program (NOP). As of 2007 over 60 countries regulate organic farming (IFOAM 2007:11). In 2005 IFOAM created the Principles of Organic Agriculture, an international guideline for certification criteria. Typically the agencies accredit certification groups rather than individual farms.

Organic production materials used in and foods are tested independently by the Organic Materials Review Institute.

Composting

Under USDA organic standards, manure must be subjected to proper thermophilic composting and allowed to reach a sterilising temperature. If raw animal manure is used, 120 days must pass before the crop is harvested if the final product comes into direct contact with the soil. For products which do not come into direct contact with soil, 90 days must pass prior to harvest.

Economics

The economics of organic farming, a subfield of agricultural economics, encompasses the entire process and effects of organic farming in terms of human society, including social costs, opportunity costs, unintended consequences, information asymmetries, and economies of scale.

Although the scope of economics is broad, agricultural economics tends to focus on maximising yields and efficiency at the farm level. Economics takes an anthropocentric approach to the value of the natural world: biodiversity, for example, is considered beneficial only to the extent that it is valued by people and increases profits.

Some entities such as the European Union subsidize organic farming, in large part because these countries want to account for the externalities of reduced water use, reduced water contamination, reduced soil erosion, reduced carbon emissions, increased biodiversity, and assorted other benefits that result from organic farming.

Traditional organic farming is labour and knowledge-intensive whereas conventional farming is capital-intensive, requiring more energy and manufactured inputs.

Organic farmers in California have cited marketing as their greatest obstacle.

Geographic Producer Distribution

The markets for organic products are strongest in North America and Europe, which as of 2001 are estimated to have $6 and $8 billion respectively of the $20 billion global market (Lotter 2003:6). As of 2007 Australasia has 39% of the total organic farmland, including Australia's 1,180,000 hectares (2,900,000 acres) but 97 percent of this land is sprawling rangeland (2007:35).

US sales are 20x as much. (2003:7). Europe farms 23 percent of global organic farmland (6.9 million hectares), followed by Latin America with 19 percent (5.8 million hectares). Asia has 9.5 percent while North America has 7.2 percent. Africa has 3 percent.

Besides Australia, the countries with the most organic farmland are Argentina (3.1 million hectares), China (2.3 million hectares), and the United States (1.6 million hectares).

Much of Argentina's organic farmland is pasture, like that of Australia (2007:42). Italy, Spain, Germany, Brazil (the world's largest agricultural exporter), Uruguay, and the UK follow the United States in the amount of organic land (2007:26).

Growth

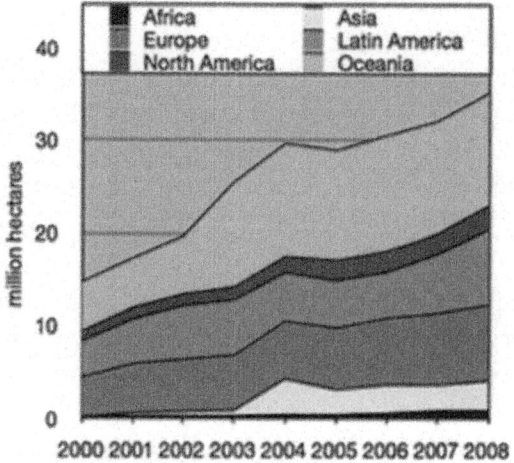

Figure: Organic farmland by world region (2000-2008)

As of 2001, the estimated market value of certified organic products was estimated to be $20 billion. By 2002 this was $23 billion and by 2007 more than $46 billion.

In recent years both Europe (2007: 7.8 million hectares, European Union: 7.2 million hectares) and North America (2007: 2.2 million hectares) have experienced strong growth in organic farmland. In the EU it grew by 21% in the period 2005 to 2008. However, this growth has occurred under different conditions. While the European Union has shifted agricultural subsidies to organic farmers due to perceived environmental benefits, the United States has not, continuing to subsidize some but not all traditional commercial crops, such as corn and sugar. As a result of this policy difference, as of 2008 4.1% percent of European Union farmland was organically managed compared to the 0.6 percent in the U.S.

IFOAM's most recent edition of *The World of Organic Agriculture: Statistics and Emerging Trends 2009* lists the countries which had the most hectares in 2007. The country with the most organic land is Australia with more than 12 million hectares, followed by Argentina, Brazil and the US. In total 32.2 million hectares were under organic management in 2007. For 1999 11 million hectares of organically managed land are reported.

As organic farming becomes a major commercial force in agriculture, it is likely to gain increasing impact on national agricultural policies and confront some of the scaling challenges faced by conventional agriculture.

Productivity and Profitability

Various studies find that versus conventional agriculture, organic crops yielded 91%, or 95-100%, along with 50% lower expenditure on fertilizer and energy, and 97% less pesticides, or 100% for corn and soybean, consuming less energy and zero pesticides. The results were attributed to lower yields in average and good years but higher yields during drought years.

A 2007 study compiling research from 293 different comparisons into a single study to assess the overall efficiency of the two agricultural systems has concluded that

> ...organic methods could produce enough food on a global per capita basis to sustain the current human population, and potentially an even larger population, without increasing the agricultural land base.

Converted organic farms have lower pre-harvest yields than their conventional counterparts in developed countries (92%) but higher than their low-intensity counterparts in developing countries (132%). This is due to relatively lower adoption of fertilizers and pesticides in the developing world compared to the intensive farming of the developed world.

Organic farms withstand severe weather conditions better than conventional farms, sometimes yielding 70-90% more than conventional farms during droughts.

Organic farms are more profitable in the drier states of the United States, likely due to their superior drought performance. Organic farms survive hurricane damage much better, retaining 20 to 40% more topsoil and smaller economic losses at highly significant levels than their neighbours.

Contrary to widespread belief, organic farming can build up soil organic matter better than conventional no-till farming, which suggests long-term yield benefits from organic farming. An 18-year study of organic methods on nutrient-depleted soil, concluded that conventional methods were superior for soil fertility and yield in a cold-temperate climate, arguing that much of the benefits from organic farming are derived from imported materials which could not be regarded as "self-sustaining".

Profitability

The decreased cost of synthetic fertilizer and pesticide inputs, along with the higher prices that consumers pay for organic produce, contribute to increased profits. Organic farms have been consistently found to be as or more profitable than conventional farms. Without the price premium, profitability is mixed. Organic production was more profitable in Wisconsin, given price premiums.

Sustainability (African Case)

In 2008 the United Nations Environmental Programme (UNEP) and the United Nations Conference on Trade and Development (UNCTAD) stated that "organic agriculture can be more conducive to food security in Africa than most conventional production systems, and that it is more likely to be sustainable in the long-term" and that "yields had more than doubled where organic, or near-organic practices had been used" and that soil fertility and drought resistance improved.

Employment Impact

Organic methods often require more labour than traditional farming, therefore it provides rural jobs.

Externalities

Agriculture imposes negative externalities (uncompensated costs) upon society through land and other resource use, biodiversity loss, erosion, pesticides, nutrient runoff, water usage, subsidy payments and assorted other problems. Positive externalities include self-reliance, entrepreneurship, respect for nature, and air quality. Organic methods reduce some of these costs. In 2000 uncompensated costs for 1996 reached 2,343 million British pounds or 208 pounds per hectare. In 2005 in the USA concluded that cropland costs the economy approximately 5 to 16 billion dollars ($30 to $96 per hectare), while livestock production costs 714 million dollars. Both studies recommended reducing externalities. The 2000 review included reported pesticide poisonings but did not include speculative chronic health effects of pesticides, and the 2004 review relied on a 1992 estimate of the total impact of pesticides.

It has been proposed that organic agriculture can reduce the level of some negative externalities from (conventional) agriculture. Whether the benefits are private or public depends upon the division of property rights.

Pesticides

Figure: A sign outside of an organic apple orchard in Pateros, Washington reminding orchardists not to spray pesticides on these trees.

Most organic farms largely avoid pesticides as opposed to conventional farms. Some pesticides damage the environment or with direct exposure, human health. Children exposed to pesticides are of special concern. According to the National Academy of Sciences:

> *"A fundamental maxim of pediatric medicine is that children are not 'little adults.' Profound differences exist between children and adults. Infants and children are growing and developing. Their metabolic rates are more rapid than those of adults. There are differences in their ability to activate, detoxify, and excrete xenobiotic compounds. All these differences can affect the toxicity of pesticides in infants and children, and for these reasons the toxicity of pesticides is frequently different in children and adults."*

The five main pesticides used in organic farming are Bt (a bacterial toxin), pyrethrum, rotenone, copper and sulphur. Fewer than 10% of organic vegetable farmers acknowledge using these pesticides regularly; 5.3% of vegetable growers will admit rotenone use; while 1.7% admit pyrethrum use (Lotter 2003:26). Reduction and elimination of chemical pesticide use is technically challenging. Organic pesticides often complement other pest control strategies.

Ecological concerns primarily focus around pesticide use, as 16% of the world's pesticides are used in the production of cotton.

Runoff is one of the most damaging effects of pesticide use. The USDA Natural Resources Conservation Service tracks the environmental effects of water contamination and concluded, "the Nation's pesticide policies during the last twenty six years have succeeded in reducing overall environmental risk, in spite of slight increases in area planted and weight of pesticides applied. Nevertheless, there are still areas of the country where there is no evidence of progress, and areas where risk levels for protection of drinking water, fish, algae and crustaceans remain high".

Food Quality and Safety

Many studies have examined the relative quality and safety benefits of organic and conventional agricultural techniques. The results are diverse. Some find no significant differences. Others disagree. An example of the "no differences" school stated:

No evidence of a difference in content of nutrients and other substances between organically and conventionally produced crops and livestock products was detected for the majority of nutrients assessed in this review suggesting that organically and conventionally produced crops and livestock products are broadly comparable in their nutrient content... There is no good evidence that increased dietary intake, of the nutrients identified in this review to be present in larger amounts in organically than in conventionally produced crops and livestock products, would be of benefit to individuals consuming a normal varied diet, and it is therefore unlikely that these differences in nutrient content are relevant to consumer health.

However, they also found that statistically significant differences between the composition of organic and conventional food were present for a few substances.

"Organic products stand out as having higher levels of secondary plant compounds and vitamin C". Organic kiwifruit had more antioxidants.

A review of potential health effects analysed eleven articles, concluding, "because of the limited and highly variable data available, and concerns over the reliability of some reported findings, there is currently no evidence of a health benefit from consuming organic compared to conventionally produced foodstuffs. It should be noted that this conclusion relates to the evidence base currently available on the nutrient content of foodstuffs, which contains limitations in the design and in the comparability of studies."

Individual studies have considered a variety of possible impacts, including pesticide residues. Pesticide residues present a second channel for health effects. Comments include, "Organic fruits and vegetables can be expected to contain fewer agrochemical residues than conventionally grown alternatives; yet, "the significance of this difference is questionable" and "It is intuitive to assume that children whose diets consist of organic food items would have a lower probability of neurologic health risks", and pesticide exposure brought an increased risk for ADHD in one study.

Nitrate concentrations may be less, but the health impact of nitrates is debated. Lack of data has limited research on the health effects of natural plant pesticides and bacterial pathogens. Consumption of organic milk was associated with a decrease in risk for eczema, although no comparable benefit was found for organic fruits, vegetables, or meat.

The higher cost of organic food (ranging from 45 to 200%) could inhibit consumption of the recommended 5 servings per day of vegetables and fruits, which improve health and reduce cancer regardless of their source.

Clothing Quality and Safety

Recently, organic clothing has become widely available. Although many consumers of organic clothing merely dislike synthetic chemicals, a significant portion of the organic clothing market comes from those suffering from Multiple Chemical Sensitivity, a chronic medical condition characterised by symptoms that the affected person says are adverse effects from exposure to low levels of chemicals.

Soil Conservation

In *Dirt: The Erosion of Civilizations*, geomorphologist David Montgomery outlines a coming crisis from soil erosion. Agriculture relies on roughly one metre of topsoil, and that is being depleted ten times faster than it is being replaced. No-till farming, which some claim depends upon pesticides, is one way to minimise erosion. However, a recent study by the USDA's Agricultural Research Service has found that manure applications in tilled organic farming are better at building up the soil than no-till.

Climate Change

Organic agriculture emphasizes closed nutrient cycles, biodiversity, and effective soil management providing the capacity to mitigate and even reverse the effects of climate change. Organic agriculture can decrease fossil fuel emissions and, like any well managed agricultural system, sequesters carbon in the soil. The elimination of synthetic nitrogen in organic systems decreases fossil fuel consumption by 33 percent and carbon sequestration takes CO_2 out of the atmosphere by putting it in the soil in the form of organic matter which is often lost in conventionally managed soils. Carbon sequestration occurs at especially high levels in organic no-till managed soil.

Agriculture has been undervalued and underestimated as a means to combat global climate change. Soil carbon data show that regenerative organic agricultural practices are among the most effective strategies for mitigating CO_2 emissions.

Nutrient Leaching

Excess nutrients in lakes, rivers, and groundwater can cause algal blooms, eutrophication, and subsequent dead zones. In addition, nitrates are harmful to aquatic organisms by themselves. The main contributor to this pollution is nitrate fertilizers whose use is expected to "double or almost triple by 2050". Organically fertilizing fields "significantly [reduces] harmful nitrate leaching" over conventionally fertilized fields: "annual nitrate leaching was 4.4-5.6 times higher in conventional plots than organic plots".

The large dead zone in the Gulf of Mexico is caused in large part by agricultural runoff: a combination of fertilizer and livestock manure. Over half of the nitrogen released into the Gulf comes from agriculture. This increases costs for fishermen, as they must travel far from the coast to find fish.

Nitrogen leaching into the Danube River was substantially lower among organic farms. The resulting externalities could be neutralised by charging 1 euro per kg of released nitrogen.

Agricultural runoff and algae blooms are strongly linked in California.

Biodiversity

A wide range of organisms benefit from organic farming, but it is unclear whether organic methods confer greater benefits than conventional integrated agri-environmental programs. Nearly all non-crop, naturally occurring species observed in comparative farm land practice studies show a preference for organic farming both by abundance and diversity. An average of 30% more species inhabit organic farms. Birds, butterflies, soil microbes, beetles, earthworms, spiders, vegetation, and mammals are particularly affected. Lack of herbicides and pesticides improve biodiversity fitness and population density. Many weed species attract beneficial insects that improve soil qualities and forage on weed pests. Soil-bound organisms often benefit because of increased bacteria populations due to natural fertilizer such as manure, while experiencing reduced intake of herbicides and pesticides. Increased biodiversity, especially from beneficial soil microbes and mycorrhizae have been proposed as an explanation for the high yields experienced by some organic plots, especially in light of the differences seen in a 21-year comparison of organic and control fields.

Biodiversity from organic farming provides capital to humans. Species found in organic farms enhance sustainability by reducing human input (e.g., fertilizers, pesticides). Farmers that produce with organic methods reduce risk of poor yields by promoting biodiversity. Common game birds such as the ring-necked pheasant and the northern bobwhite often reside in agriculture landscapes, and benefit recreational hunters.

Sales and Marketing

Most sales are concentrated in developed nations. These products are what economists call credence goods in that they rely on uncertain certification. Interest in organic products dropped between 2006 and 2008, and 42% of Americans polled don't trust organic produce. 69% of Americans claim to occasionally buy organic products, down from 73% in 2005. One theory was that consumers were substituting "local" produce for "organic" produce.

Distributors

In the United States, 75% of organic farms are smaller than 2.5 hectares. In California 2% of the farms account for over half of sales. Small farms join together in cooperatives such as Organic Valley, Inc. to market their goods more effectively.

Most small cooperative distributors have merged or were acquired by large multinationals such as General Mills, Heinz, ConAgra, Kellogg, and others. In 1982 there were 28 consumer cooperative distributors, but as of 2007 only 3 remained. This consolidation has raised concerns among consumers and journalists of potential fraud and degradation in standards. Most sell their organic products through subsidiaries, under other labels.

Organic foods also can be a niche in developing nations. It would provide more money and a better opportunity to compete internationally with the huge distributors. Organic prices are much more stable than conventional foods, and the small farms can still compete and have similar prices with the much larger farms that usually take all of the profits.

Farmers' Markets

Price premiums are important for the profitability of small organic farmers. Farmers selling directly to consumers at farmers' markets have continued to achieve these higher returns. In the United States the number of farmers' markets tripled from 1,755 in 1994 to 5,274 in 2009.

Capacity Building

Organic agriculture can contribute to ecologically sustainable, socio-economic development, especially in poorer countries. The application of organic principles enables employment of local resources (e.g., local seed varieties, manure, etc.) and therefore cost-effectiveness. Local and international markets for organic products show tremendous growth prospects and offer creative producers and exporters excellent opportunities to improve their income and living conditions.

Organic agriculture is knowledge intensive. Globally, capacity building efforts are underway, including localised training material, to limited effect. As of 2007, the International Federation of Organic Agriculture Movements hosted more than 170 free manuals and 75 training opportunities online.

Controversy

Norman Borlaug (father of the "Green Revolution" and a Nobel Peace Prize laureate), Prof A. Trewavas and other critics contested the notion that organic agricultural systems are more friendly to the environment and more sustainable than conventional farming systems. Borlaug asserts that organic farming practices can at most feed 4 billion people, after expanding cropland dramatically and destroying ecosystems in the process. The Danish Environmental Protection

Agency estimated that phasing out all pesticides would result in an overall yield reduction of about 25%. Environmental and health effects were assumed but hard to assess.

In contrast, the UN Environmental Programme concluded that organic methods greatly increase yields in Africa. A review of over two hundred crop comparisons argued that organic farming could produce enough food to sustain the current human population and that the difference in yields between organic and non-organic methods were small, with non-organic methods yielding slightly more in developed areas and organic methods yielding slightly more in developing areas.

That analysis has been criticised by Alex Avery of the Hudson Institute, who contends that the review claimed many non-organic studies to be organic, misreported organic yields, made false comparisons between yields of organic and non-organic studies which were not comparable, counted high organic yields several times by citing different papers which referenced the same data, and gave equal weight to studies from sources which were not impartial.

The Centre for Disease Control repudiated a claim by Avery's father, Dennis Avery (also at Hudson) that the risk of E. coli infection was eight times higher when eating organic food. (Avery had cited CDC as a source.) Avery had included problems stemming from non-organic unpasteurised juice in his calculations. Epidemiologists traced the 2011 E. coli O104:H4 outbreak - which caused over 3,900 cases and 52 deaths - to an organic farm in Bienenbüttel in Germany.

Loss of Soil Organic Matter and its Restoration

Centuries before there was any science that acquainted people with the intricacies of plant nutrition, decaying organic matter, as in manure or other forms, was recognised as an effective agent in the nourishment of plants. The high productivity of most virgin soils has always been associated with their high content of organic matter, and the decrease in the supply with cultivation has generally been paralleled by a corresponding decrease in productivity. Even though we can now feed plants on diets that produce excellent growth without the use of any soil whatever, yet the decaying remains of preceding plant generations, resolved by bacterial wrecking crews into simpler, varied nutrients for rebuilding into new generations, must still be the most effective basis for extensive crop production by farmers. Soil organic matter is one of our most important national resources; its unwise exploitation has been devastating; and it must be given its proper rank in any conservation policy as one of the major factors affecting the levels of crop production in the future.

The stock of organic matter in the virgin soils taken over by the homesteading pioneers was a heritage from an extensive past. Its accumulation in our northern soils began with the recession of the last glacier, possibly some 25,000 years ago, and continued long enough to ripen the residues into compounds that were ready to be used quickly by growing plants.

With the departure of the ice sheet and the consequent general rise in temperature, the glacial residue of pulverized rock offered minerals in solution for plant growth. As the plants found nitrogen to combine with these minerals, they grew, died, and began to accumulate in the soil.

Then, as the rate of rock weathering increased, bringing a larger supply of soluble minerals, the accumulation of plant remains became correspondingly larger. Finally, when the rocks were more completely weathered so that they provided less mineral stock, or very little, an equilibrium point was reached at which the accumulated organic matter held in combination most of the minerals that could be turned into soluble forms. Thereafter the supply of soluble minerals became a limiting factor in plant growth.

Wherever there was poor drainage and limited aeration of the sod cover, or where there were heavily wooded soils of relatively level, glaciated topography, more complete simplification of this accumulated store of plant nutrients was very slow.

In other words, the organic matter that held the major stock of previously mobile nitrogen and minerals now kept these essentials stored in compounds not simple enough for prompt consumption by growing plants. This represented a very large supply of nutrients not far from the condition in which growing plants could use them. Unable to decay completely or to accumulate much more, they were poised as it were for rapid conversion, when a slight change in conditions occurred, into forms of maximum utility for plant growth.

But with the removal of water through furrows, ditches, and tiles, and the aeration of the soil by cultivation, what the pioneers did in effect was to fan the former simmering fires of acidification and preservation into a blaze of bacterial oxidation and more complete combustion. The combustion of the accumulated organic matter began to take place at a rate far greater than its annual accumulation. Along with the increased rate of destruction of the supply accumulated from the past, the removal of crops lessened the chance for annual additions. The age-old process was reversed and the supply of organic matter in the soil began to decrease instead of accumulating.

Fuel for the Plant-Food Production Factory

Organic matter may well be considered as fuel for bacterial fires in the soil, which operates as a factory producing plant nutrients. The organic matter is burned to carbon dioxide, ash, and other residues. This provides carbonic acid in the soil water, and the solvent effect of this acidified water on calcium, potassium, magnesium, phosphates, and other minerals in rock form is many hundreds of times greater than that of rain water.

At the same time the complex constituents of the organic matter are simplified, and nitrogen in the ammonia is released and converted into the nitrate form. This, very briefly, is the complicated process of decomposition, from which carbon dioxide results as the major simplified end product, together with a host of others in smaller amounts. This gas is released in such large quantities from the soil that the supply in the atmosphere over the earth is maintained at a constant amount.

Decomposition by micro-organisms within the soil is the reverse of the process represented by plant growth above the soil. Growing plants, using the energy of the sun, synthesize carbon, nitrogen, and all other elements into complex compounds. The energy stored up in these compounds is then used more or less completely by the microorganisms whose activity within the soil makes nutrients available for a new generation of plants. Organic matter thus supplies the "life of the Soil" in the strictest sense.

When measured in terms of carbon dioxide output, the soil is a live, active body. An acre of the better Corn Belt soil in Iowa *(365)* or northern Illinois, for example, exhales more than 25 times as much of this gas per day as does an adult man at work. Such a soil area burns carbon at a rate equivalent to 1.6 pounds of a good grade of soft coal per hour.

The heat equivalent evolved in the same time would convert more than 17 pounds of water to steam under 100 pounds pressure. A 40-acre cornfield during the warmer portion of a July day is burning organic matter in the soil with an energy output equivalent to that of a 40-horsepower steam engine; every acre, in other words, may be roughly pictured as a factory using the equivalent of 1 horsepower. Organic matter is the source of the power without which the plant-food elements could not be changed to usable forms.

Supply of Virgin Soil Organic Matter Decreasing

The depletion of the supply of organic matter by cultivation is well illustrated by the report of a study made by Jenny *(185)* in central Missouri in which an undisturbed virgin prairie soil was compared to an adjoining field cropped to corn, wheat, and oats for 60 years without the addition of manure or fertilizer. No erosion had taken place, yet 38 percent of the organic matter represented by the virgin soil had been lost during that period because of cultivation. As a consequence of this loss in organic matter, the soil structure was modified to an extent that might be represented by reducing the number of granules that were the size of particles of sand by 11 percent and increasing the number that were the size of clay particles by 5.5 percent. The loss of organic matter represents soil compaction, which hampers the circulation of air and water and hinders tillage operations at the same time that the function of the soil in plant nutrition is disturbed. Thus in but 60 years, more than one-third of the organic matter, representing centuries of accumulation, was destroyed and the efficiency of the soil for crop production was reduced.

Resulting Loss in Nitrate Nitrogen

Soil organic matter is the source of nitrogen. In the later stages of decay of most kinds of organic matter, nitrogen is liberated as ammonia and subsequently converted into the soluble or nitrate form. The level of crop production is often dependent on the capacity of the soil to produce and accumulate this form of readily usable nitrogen. We can thus measure the activity that goes on in changing organic matter by measuring the nitrates. It is extremely desirable that this change be active and that high levels of nitrate be provided in the soil during the growing season.

A study of the nitrate levels under corn in a Missouri silt loam during 13 years reveals a gradual decline in the production of nitrates *(5)*. During the first 5 years of the test this soil increased its nitrates in the spring to the maximum of more than 20 pounds per acre as early as May. During a similar period only 2 years later, this maximum had been reduced to 18 pounds, and it was not attained until June;

3 years later the maximum was less than 16 pounds, attained in July; and 3 years after that, the maximum of 13 pounds was not reached until August. During continuous cropping to corn without the addition of organic matter, the maximum nitrate accumulation dropped to 65 percent of that in the initial period, when the land had been in sod for some time. In other words, though this soil had been in corn continuously for only 13 years—which might seem equal to 52 years of a 4-year rotation, with one crop of corn every 4 years— its nitrate-producing power, or its capacity to deliver this soluble plant nutrient, had been reduced by 35 percent.

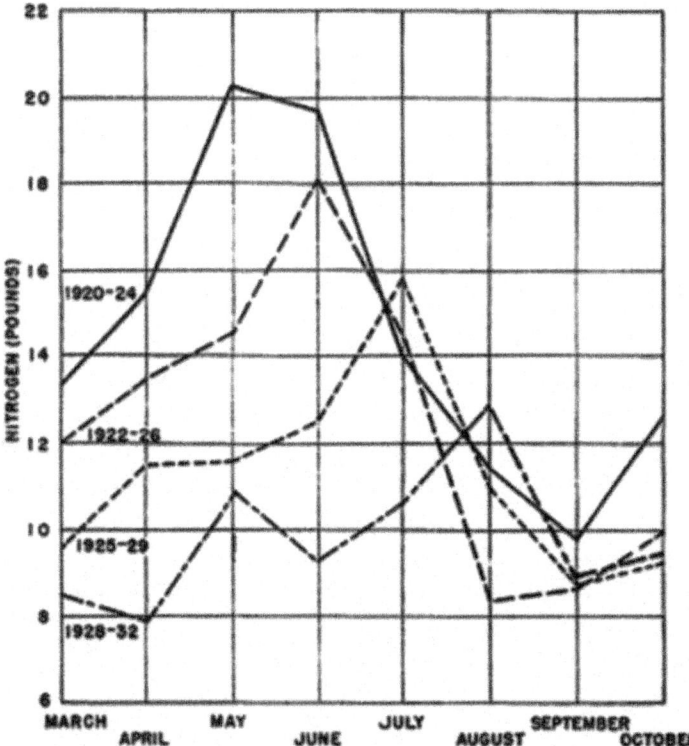

Figure: Declining seasonal levels of nitrates and later seasonal maxima with continued cropping to corn. (Averaged for different succeeding 5-year periods.)

Such pronounced exhaustion is not limited to the corn crop, which is readily associated with intensive tillage. The same thing is true in the case of wheat. Its exhausting effects measure were even greater than those of corn. The nitrate level under wheat was constantly lower than under corn for the corresponding period.

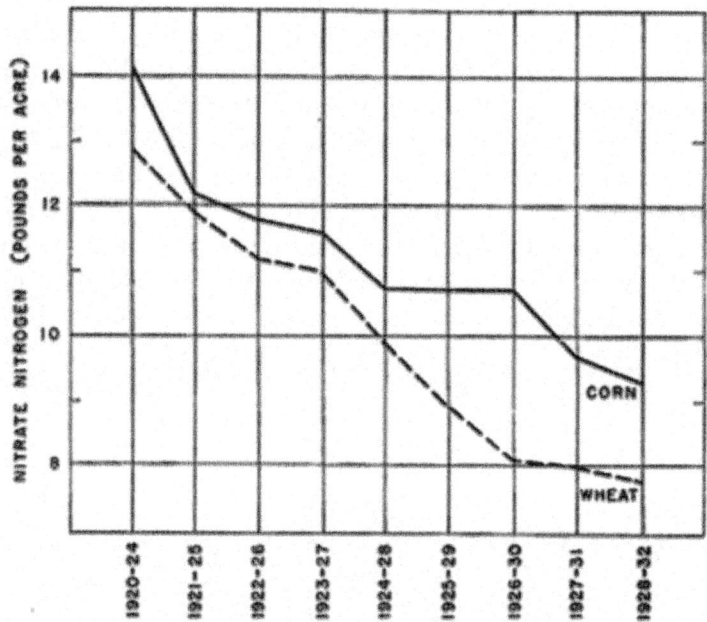

Figure: Declining nitrate nitrogen levels in soils in wheat and in corn as advancing 5-year season averages.

Concurrently with the foregoing measurements showing the decline of nitrates, careful chemical analyses were made of the same kind of soil nearby under fallow conditions with an annual spring plowing. The surface soil alone lost 2,300 pounds of organic matter per acre. A nearby plot in a 3-year rotation of corn, wheat, and clover, with all the crops removed during a period beginning 2 years earlier and extending 2 years longer—a total of 17 years—lost 800 pounds of organic matter.

From unpublished data supplied by M. F. Miller Missouri Agricultural Experiment Station. Regardless of the presence or absence of a crop, the failure to add organic matter and regular tillage of the soil mean a depletion of the original stock of organic matter at a very significant rate, even where there is no erosion. Where erosion removes the body of the surface soil itself, the rate of depletion is much greater.

Lower Crop Yields and Land Values

In addition to carrying nitrogen, the nutrient demanded in largest amount by plants, soil organic matter either supplies a major portion of the mineral elements from its own composition, or it functions to move them out of their insoluble, useless forms in the rock minerals into active forms within the colloidal clay. Organic matter itself is predominantly of a colloidal form resembling that of clay, which is the main chemically active fraction of the soil.

But it is about five times as effective as the clay in nutrient exchanges. Nitrogen, as the largest single item in plant growth, has been found to control crop-production levels, so that in the Corn Belt crop yields roughly parallel the content of organic matter in the soil *(184)*. On a Missouri soil with less nitrogen than that corresponding to 2 percent of organic matter (40,000 pounds of organic matter per acre of plowed surface soil) an average yield of 20 bushels of corn per acre can hardly be expected.

For yields approaching 40 bushels, roughly double the amount of organic matter is required. With declining organic matter go declining corn yields and therefore lower earnings on the farmers investment. Thus the stock of organic matter in the soil, particularly as measured by nitrogen, is a rough index of land value when applied to soils under comparable conditions. According to studies in Missouri, for example, the lower the content of organic matter of upland soil, the lower the average market value of the land.

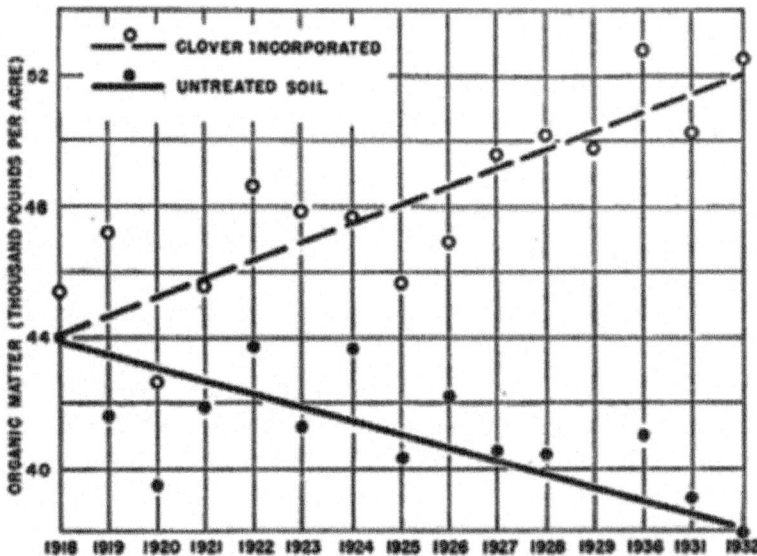

Figure: Decrease of organic-matter content in a fallow, untreated soil in contrast to the gain in soil treated with 2 1/2 tons of red clover annually, representing over 500 pounds annual increase in organic matter per acre.

Problem of Maintaining a Liberal Supply of Soil Organic Matter

Though the rapid depletion in the Corn Belt, for example, of the soil organic matter and soil fertility in the pioneer period of a hundred years may be alarming, there is consolation in the fact that this high rate of depletion will not continue. As is true for all biochemical processes, the early rate of consumption is rapid, which gives a sudden decrease. Then the rate of consumption falls off, so that the loss in the second period will perhaps be less than half that in the first.

In the third stage the loss will possibly be half that of the second. Long-continued experiments, accompanied by soil analyses, prove that the organic-matter content of a soil will reach a fixed level characteristic of the surrounding climatic conditions. After a period of heavy loss, then, we may expect a fairly constant level during a long period of continued cultivation. This situation is well illustrated by the decline in nitrogen content, Soil Nitrogen (p.369). In other words, we may anticipate a further decline in productivity from the present relatively high levels, followed by a more constant level, which will be proportionate to the lower content of organic matter, determined by the environment in each particular region.

Maintaining Versus Increasing the Organic Matter

The following questions naturally arise: What should be the content of organic matter in a soil? Should the present level be raised or merely maintained economically? These are questions of decided significance in determining policies in soil management.

Attempting to hoard as much organic matter as possible in the soil, like a miser hoarding gold, is not the correct answer. Organic matter functions mainly as it is decayed and destroyed. Its value lies in its dynamic nature. A soil is more productive as more organic matter is regularly destroyed and its simpler constituents made usable during the growing season.

Its mere presence in the soil is of value during certain stages of decay, when it influences soil structure and water relations and when it functions in holding plant food in readily available form much more effectively than does any mineral fraction of the soil. The objective should be to have a steady supply of organic matter undergoing these processes for the benefit of the growing crop. Up to the present, the policy—if it can be called a policy—has been to exhaust the supply, rather than to maintain it by regular additions according to the demands of the crops produced or the soil fertility removed. To continue very long with this practice will mean a further sharp decline in crop yields.

The level of organic-matter content to be maintained is not the same for all regions. It varies according to climate. Professor Jenny in his studies of virgin organic matter of soils *(184)* has pointed out that—within regions of similar moisture conditions, the organic matter content of upland, terrace, and bottomland soil, including both prairie and timber vegetation, decreases from north to south. For each fall of 10° C. (18° F.) in annual temperature, the average organic matter content of the soil increases two or three times, provided the precipitation-evaporation ratio is kept constant.

Thus from south to north the level of organic matter in the soil becomes naturally higher. In the northern section of the Temperate Zone with its moderate rate of vegetative growth and moderate production of organic matter, the longer periods of lower temperature lessen decay and increase accumulation by carry-over from season to season. In the southern section, even though the growing season is longer and produces more vegetation, yet there is also a longer season for decay, and it proceeds at a much more rapid rate. Because the rate of decay doubles and trebles for every rise of 10° C. (18° F.) in temperature, the destruction of organic matter is more complete and there is little accumulation. Its nature, particularly its composition, is also different. It shows a narrower carbon-nitrogen ratio *(184)* and a greater resistance to further simplification.

The level of organic matter in the soil of the temperate regions rises with lower annual temperatures, and also with increased moisture. The level is also higher in grasslands than in timbered soils under equal moisture conditions. The same amount of moisture in the North with its lower temperature is more effective in bringing about an increase of soil organic matter than in the South with its higher temperature. Hence sod crops are more effective restorers of organic matter in the northern than in the southern part of the North Temperate Zone. The climate of the region must be considered in determining the level of organic matter to be maintained in the soil. Changes in altitude must also be considered insofar as these correspond to climatic variations.

In northern Missouri, for example, virgin soils are in a condition of natural equilibrium at an organic-matter content of 3.54 percent;

in southern Missouri at 2.20 percent; in southern Minnesota at 4.44 percent; and in Arkansas at 1.96 percent. In terms of pounds per plowed acre, the figures are: For southern Minnesota, 88,800 pounds; northern Missouri, 70,800 pounds; southern Missouri, 44,000 pounds; and Arkansas, 39,200 pounds.

These figures represent the natural equilibrium between the production of organic matter by native vegetation and its destruction by micro-organisms. The balance figure is determined in the main by the temperature-rainfall combination, or climate. It would be folly, according to these data, for the farmer in Arkansas to attempt to increase organic matter in his soil to the level common in the soil tilled by the Minnesota farmer. Likewise the problem of increasing the organic matter will be simpler for the farmer in the North, where even with the same amount of moisture, the lower temperature is influential in preserving more of the organic matter added to the soil.

Cultivation of the soil and extended periods without a vegetative cover decrease the content of organic matter below that considered natural, or virgin, for the locality. The degree of exhaustion of organic matter to levels below the virgin stock represents the possibilities of improvement. But these possibilities also are affected by climate. In the northern sections both temperature and moisture conditions are favourable to restoration, and the growing of legumes and the addition of green manure are very effective in this direction, as experimental results demonstrate. Farther south, restoration is more difficult, and it may even be impossible to restore the organic matter profitably and permanently to levels even approaching virgin conditions. However, the longer growing season permits two crops a year, one of which may be a legume for green manure, and this makes it possible to provide organic matter and a turnover of nitrogen regularly even when the level cannot be raised.

We are confronted, then, by three facts: (1) The stock of organic matter in the soil is being exhausted at an alarming rate; (2) the exhaustion is still in its early stages in some of the more recently developed agricultural areas; and (3) there are no climatic handicaps that prohibit restoration. These facts mean corresponding—and inescapable—responsibilities. The Nation should be made aware of the rapid rate at which the organic matter in the soil is being exhausted. Farm-management practices should be adopted that will at least maintain, and in as many cases as possible even increase, the supply of this natural resource in the soil. The maintenance of soil organic matter might well be considered a national responsibility.

Interrelation of Soil Organic Matter with Nitrogen and Minerals

At first thought, the problem of restoring soil organic matter may not seem difficult according to simple mathematical calculations. If a soil in virgin condition contained 44,000 pounds of organic matter per acre and 35 percent of this has been exhausted during 60 years of cultivation, the apparently simple solution would be to add 15,400 pounds of dry material to the soil, or an amount of organic matter equivalent to the weight lost. The addition of the equivalent of some 7 3/4 tons of dry matter in the form of manure, legumes, straw, and other farm-waste products might seem to be a satisfactory solution. But the virgin organic matter that has been lost was very different in nature and effects from the material considered to replace it. In kind and composition, the organic matter used for restoration should be as close as possible to that which was lost, at least in terms of effective results.

Building Soil Organic Matter Largely a Nitrogen Problem

Soil bacteria, the agents of decomposition, use carbon mainly as fuel and nitrogen as building material for their bodies and for the production of the intricate organic compounds that result from their activity. Fresh organic matter is characterised as a rule by a large amount of carbon in relation to nitrogen. It has a wide carbon-nitrogen ratio, in other words; or so far as the bacteria are concerned, a wide ratio of fuel to building material. Such fresh material—straw, for example—may have a ratio that is too wide, so that it decomposes very slowly. If the ratio is less wide, decomposition may be more actively carried on. The carbon will then be rapidly used up as fuel while the nitrogen is held or treasured without appreciable loss.

Thus when decay has proceeded to the point where the carbon-nitrogen ratio is significantly decreased, a residue of a more stable nature is produced. Thereafter the carbon-nitrogen ratio is narrower and remains more constant. This corresponds more nearly to the condition that holds in the case of the organic matter in virgin soils. Its further decay, which is slow because of the relatively low level of carbon, liberates nitrogen in place of storing or preserving it. Because of its high carbon content, the decomposition of fresh organic matter requires additional soluble nitrogen to be used as building material by the micro-organisms, which obtain it from the soil, often exhausting the supply to a degree that is damaging to a growing crop. The amount of increase in organic material corresponds, in the main, to the amount of nitrogen available. The extra carbon in the fresh material is lost from the soil. Thus when soils are given straws, fodders, and similar crop residues of low nitrogen content, only small increases in soil organic matter can result—in the main, only as large as the added nitrogen will permit. Many tons of common farm residues and wastes per acre are needed to produce a single additional ton of organic matter in the soil.

The restoration of soil organic matter, then, is a problem of increasing the nitrogen level or of using nitrogen as a means of holding the carbon and other materials. This is the basic principle behind the use of legumes as green manures. In building up the organic content of the soil itself, it will often be desirable to use legumes and grasses rather than to add organic matter, such as straw and compost, directly. If legumes and grasses are to be successfully grown on many of the soils of the humid regions of this country it will be necessary, first, to properly fertilize and lime the soil. Legumes use nitrogen from the air instead of the soil, and thus serve to increase the amount in the soil when their own remains are added to it. Commercial nitrogen used as treatment on straw for the production of artificial manure in compost piles, or when plowing under straw in the field after the combine, may be considered in the same category. Small amounts of added nitrogen may in this way make possible the use of large amounts of carbonaceous matter in restoring the soil. Thus the European farmer first "makes" his manure by composting the fresh straw-dung mixture from the barn and then treats it intermittently with the nitrogen-bearing liquid manure or urine from the same source and the nitrogen-rich leachings from the manure pit. He does not consider the fresh, strawy barn waste manure in the strictest sense until the surplus carbon has been removed through the heating process, and the less active manure compounds become similar to those of the soil organic matter. In a similar way, it should be understood that the soil organic matter can be "made" or built up only as the nitrogen supply is raised and combined with carbonaceous material in a more narrow ratio.

It is only under conditions of this kind that beneficial effects on crops may be expected through further decomposition. The manure making of the Old World farmer turns the miscellaneous straw-dung-urine mixture, of highly variable value, into a standardised fertilizer for specific use. Our great variety of crop wastes—straw, cornstalks, etc.—should be used in a similar way, by adding nitrogen to bring about a proper adjustment with their excess carbon. These neglected wastes will then provide extra and valuable soil organic matter that will have beneficial rather than possibly detrimental effects on crops.

Level of Minerals in Soil Influences Organic-Matter Supply

Bacterial activity does not occur in the absence of the mineral elements, such as calcium, magnesium, potassium, phosphorus, and others. These, as well as the nitrogen, are important: Recent studies show that the rate of decomposition is reduced when the soil is deficient in these elements. In virgin soils high in organic matter, these elements also are at a high level, and are reduced in available forms as the organic matter is exhausted. A decline in one is accompanied by a decline in the other.

It has been held that calcium, for example, is instrumental in retaining the organic matter in a stable form in the soil. Though this seems doubtful in view of the fact that the addition of lime to soils hastens the rate of loss of organic matter, calcium has a decided influence on the growing crop and therefore on the amount of material it adds to the soil when turned under. It has recently been discovered that the fixation of nitrogen from the atmosphere by legumes is more effective where high levels of calcium are present in available form *(3)*. Thus, if in calcium-laden soils, excellent legume growth results and correspondingly large nitrogen additions are made, such soils may be expected to contain much organic matter. Liberal calcium supplies and liberal stocks of organic matter are inseparable. The restoration of the exhausted lime supply exerts an influence on building up the supply of organic matter in ways other than those commonly attributed to liming.

In the presence of lime (calcium) the legumes use other elements more effectively, such as phosphorus *(175)* and probably other nutrients. Thus heavier production results on soils rich in minerals, including more intensive and extensive root development—the most effective means of introducing organic matter into the soil. The presence of large supplies of both organic matter and minerals points clearly to the fact that the soils were high in the latter when the former was produced. It seems logical to ascribe causal significance to the minerals in the production of organic matter, whether or not they are effective in preserving it. If the soils that have lost their organic matter are to be restored, the loss of minerals, which has probably been fully as great, must be taken into account, and provision must be made to restore these mineral deficiencies before attempting to grow crops for the sake of adding organic matter.

How Can Soil Organic Matter Be Restored?

Conservation and restoration of soil organic matter as a national problem calls for a program of soil and farm management in which (1) needless losses are eliminated or reduced to a minimum, and (2) the stock in process of consumption is regularly maintained with attention to its possible economical increase. Experimental results indicate the steps in such a program.

First attention should be given to eliminating accelerated erosion. When, according to the long-continued soil erosion studies at the Missouri Agricultural Experiment Station *(263),* the entire plowed surface soil under continuous corn may be washed away in 50 years, it would be foolhardiness to attempt soil building by processes so slow as to make only an inch in hundreds of years.

Erosion can be eliminated, as the investigations have shown and recent extensive erosion-control experience demonstrates, by sod cover crops, reduction in the amount of tillage, and other measures. The establishment of sod crops on badly eroded land often requires proper fertilization and liming.

Sod crops have not been fully appreciated. Grasses have been the stepchildren in the American crop family. They have not been "cultivated" in the same sense as farm crops; they have been left to themselves, to grow on soils often turned over to them because depleted fertility made cereal cropping unprofitable. They have been incidental in the farm program. Consequently, they have not delivered their maximum in animal production and have often been very inefficient feed. Land in grass was considered idle and checked off the accounts, even if not recorded on the debit side.

The Old World, with its longer agricultural experience, shows that the lands still in good production today are those occupied by sod crops regularly for a large part of the time, where clean, or summer, cultivation has been reduced to a minimum. In France and England only slightly more than one-fourth of the cultivated soils are in clean cultivation. In Germany the figure is even less, and there are vast acreages of permanent pastures in all these countries. In the United

States the area in clean cultivation and row crops approaches one-half the cultivated land; and this in regions where the rains are of torrential nature. We may well be guided by Old World experience, which tells us that sod crops are a paramount factor in holding the soil and maintaining its productivity by their regular additions of organic matter. The tough sod slice should be more fully appreciated as an asset in terms of its organic matter rather than considered as a liability because of the high power required to plow it.

Some recent studies suggest that we have not appreciated sod crops in relation to moisture absorption and the storage of moisture in the subsoil. The beneficial effects of sods turned under for corn crops have usually been ascribed to nitrogen, when possibly the important factor has been accumulated moisture in the subsoil. Grass crops absorbed 87.4 percent of the rainfall, a 3-year rotation with one sod crop absorbed 85.5 percent, while continuous corn absorbed only 69.6 percent, according to trials extending over 14 years *(263)*. This amounted to an increased rainfall of 7.2 inches for grass and 6.4 inches for rotation as compared to continuous corn. The difference in crop yield was more significant than these figures indicate, since two-thirds of the annual rainfall came in the 6 months of the growing season, or the period when differences in rainfall mean increased yields.

Much of the extra water absorbed moves beyond the zone of consumption by the shallow grass roots and is stored there. Thus the deeper soil layer under sod, such as the third foot, carries more water than the same layer under tilled soil. Moisture studies of two such adjacent soils, no far distant from those under the erosion study cited above, are interesting from this standpoint, especially for the years 1934 and 1936, which were seasons of deficient rainfall. The moisture content as the percentage of moisture in the successive 1-foot layers to a depth of 3 feet.

Table 1. Moisture content at successive depths under sod and under cultivated soil

Year and month	First foot		Second foot		Third foot	
	Sod	Tilled	Sod	Tilled	Sod	Tilled
1934:						
	Percent	Percent	Percent	Percent		Percent
April	27.18	25.23	29.90	24.61	26.11	16.51
November	33.80	31.70	31.90	30.80	32.60	24.80

1936:

March	26.30	27.80	28.20	28.90	28.30	23.00
November	27.00	26.80	28.50	27.30	27.80	19.80
1937:						
April	32.90	28.30	30.00	28.60	30.70	23.40

The third-foot layer was much drier under the tilled soil than under the sod during all of these studies. Its recovery of moisture after rain was always delayed and its total water content never equalled that in the third foot under sod.

Though the first-foot layer under sod had a lower moisture content than that under tillage during 1 month (March 1936), in all samplings the moisture supply of the second- and third-foot layers was greater under the sod than under the tilled surface, with the most pronounced differences in the third foot, varying from 5.3 to 9.5 percent. These differences mean on the average that the third-foot layer under sod is storing the equivalent of a 1.2-inch rainfall, which it may supply to the sod crop, or to the deeper roots of the following crop, in the drier summer season. This stored moisture under sod should be considered as a factor in combating droughts.

The advantages of grass-sod crops as effective agencies for soil restoration may be summed up as follows: They do much toward guaranteeing a moisture supply for their own needs by absorbing more of the rainfall. They add a heavy root growth annually that, for native bluestem, for example, amounts to as much as 1.34 tons per surface acre-foot, according to Weaver and Harman *(453)*. Because of the annual death of part of these roots, this is a regular addition of organic matter that helps to maintain the supply. On the untilled and less violently aerated soil, where the higher moisture means lower temperatures, these conditions favour a return to the original, or virgin, stock of organic matter in the soil. At the same time, erosion is prevented both by the living grass and by the spongy surface residue accumulated above the soil from the dead plant tops of the previous season.

Sod crops are sufficiently effective in restoring soil organic matter to offset the destructive influences of clean cultivation and summer tillage. Unpublished data from studies by the Missouri Agricultural Experiment Station show clearly the destructive influence of summer fallow and, in contrast, the increase in organic matter obtained through sod crops. When sod was used in an ordinary 3-year crop rotation, with manure made from the crops returned to the soil, there was no serious decline in the content of organic matter.

Table 2.: Gains and losses in soil organic matter (in pounds per acre of surface soil) during 17 years, on areas under different systems of cropping and management *

Crop and management	Gain	Loss
	Pounds	*Pounds*
Rotation—corn, wheat, clover—all crops removed	—	800
Rotation—corn, wheat, clover—manure equivalent returned	3,200	
Rye and cowpeas—turned under as green manure	1,200	
Rye turned under—summer fallow	—	14,400
Red clover continuously—all crops removed	3,600	
Red clover continuously—all crops turned under	9,600	
Alfalfa continuously—all crops removed	10,400	
Grass sod, clipped—nothing removed	10,000	

* Unpublished data of M.F. Miller, Missouri Agricultural Experiment Station.

The provision of liberal supplies of soluble plant nutrients for profitable cereal production demands tillage and the breaking down of organic matter. The organic matter to be broken down should be provided by sod crops, particularly legumes, used regularly in the rotation. Permanent legume sods are effective agents, as this study testifies, in building up the organic matter on soil containing ample minerals—particularly when the crops are not removed. Continuous red clover sod with no crop removal increased the organic-matter content by a total of 9,600 pounds in 17 years, or an average of 564 pounds a year. In similar soil in another plot nearby that was given 2 1/2 tons of clover annually as a green manure under fallow; the annual gain in organic matter amounted to 571 pounds a year.

New Awareness and New Responsibility

American citizens are becoming conscious of the fact that loss of fertility and the depletion of organic matter in the soil are partly responsible for the menace of erosion. The first step in remedying this situation is to restore fertility by the use of lime and fertilizer. The second step is to put some lands permanently into sod crops—legumes wherever possible, and the better grasses—and to use sod more regularly in rotations on tillable cropped lands. The conservation and use of such farm wastes as crop residues and manures should be included as the third step.

If these practices are recommended as proper soil management by all agricultural agencies, their adoption by individual farmers will become so common that the rate of soil depletion will be lessened. The need for longtime investments in materials that build up the soil in organic matter and fertility should be recognised in granting credit to farmers. Both owners and tenants must accept responsibility for soil conservation and work for it cooperatively. Unearned increment, the great wealth producer of the past, should be recognised as largely responsible for the mining of soil fertility and the burning up of soil organic matter until it has reached such a low level that this source of wealth has an extremely uncertain outlook in the future. The heritage of soil fertility and organic matter that we are handing on to the next generation is not large enough to be used lavishly. Careful conservation and thrifty management will be imperative if it is to yield even a moderate income.

Soil Salinisation

Salinisation is the accumulation of soluble salts of sodium, magnesium and calcium in soil to the extent that soil fertility is severely reduced.

Salinity is the degree to which water contains dissolved salts. Salinity is usually expressed in parts per thousand or grams per thousand grams. Normal seawater has a salinity of 33 parts per thousand. This rises to 40 parts per thousand in the Red Sea.

Salinisation is the process that leads to an excessive increase of water-soluble salts in the soil. The accumulated salts include sodium, potassium, magnesium and calcium, chloride, sulphate, carbonate and bicarbonate (mainly sodium chloride and sodium sulphate). A distinction can be made between primary and secondary salinisation processes. Primary salinisation involves salt accumulation through natural processes due to a high salt content of the parent material or in groundwater. Secondary salinisation is caused by human interventions such as inappropriate irrigation practices, e.g. with salt-rich irrigation water and/or insufficient drainage.

Sodification is the process by which the exchangeable sodium (Na) content of the soil is increased. Na+ accumulates in the solid and/or liquid phases of the soil as crystallised $NaHCO_3$ or Na_2CO_3 salts (salt "effloresces"), as ions in the highly alkaline soil solution (alkalisation), or as exchangeable ions in the soil absorption complex (ESP).

Soil salinisation affects an estimated 1 to 3 million hectares in the enlarged EU, mainly in the Mediterranean countries. It is regarded as a major cause of desertification and therefore is a serious form of soil degradation. Salinisation and sodification are among the major degradation processes endangering the potential use of European soils.

Soil Compaction

SOIL COMPACTION is the term for the deterioration of soil structure (loss of soil features) by mechanistic pressure, predominantly from agricultural practices.

Definition of the Problem

Soil compaction is a form of physical degradation resulting in densification and distortion of the soil where biological activity, porosity and permeability are reduced, strength is increased and soil structure partly destroyed. Compaction can reduce water infiltration capacity and increase erosion risk by accelerating run-off. The compaction process can be initiated by wheels, tracks, rollers or by the passage of animals.

Some soils are naturally compacted, strongly cemented or have a thin topsoil layer on rock subsoil. Soils can vary from being sufficiently strong to resist all likely applied loads to being so weak that they are compacted by even light loads.

In arable land with annual ploughing, both topsoil and subsoil compaction is possible. A feature of compacted soils is the formation of a pan-layer, caused by the tractor tyres driving directly on the subsoil during ploughing (above).

The pan-layer is less permeable for roots, water and oxygen than the soil below and is a bottleneck for the function of the subsoil. Unlike topsoil, the subsoil is not loosened annually, compaction becomes cumulative and over time, a homogeneous compacted layer is created.

Driving heavy tractors on the subsoil during ploughing and harvesting is a major cause of subsoil compaction. The picture on the left clearly shows how the wheels on one side of the tractor are driven in the plough furrow and press directly on the subsoil (JJHVDA).

The Impact

Large spaces in soils are known as macro pores and are created by plant roots, burrowing creatures and shrinkage caused by the drying of wet soil. These macro pores are usually continuous and form "highways" for air and water to travel deep into the soil. To an extent, continuous macro pores determine the soil's physical and soil biological quality. Macro pores are the most vulnerable pores to soil compaction.

The loss of macro porosity and pore continuity reduces strongly the ability of the soil to conduct water and air.

- Reduced infiltration capacity results in surface run-off, leading eventually to flooding, erosion and transport of nutrients and agrochemicals to open water.
- A poor aeration of the soil reduces plant growth and induces loss of soil nitrogen and production of greenhouse gases through denitrification in an aerobic sites.

Deformation of soil aggregates and higher bulk density increase the strength of the soil. This limits root growth which can result in a higher vulnerability of the crop to diseases. Subsoil compaction is a hidden form of soil degradation that can affect all the agricultural areas and results in gradually decreasing yields and gradually increasing problems with waterlogging.

In the image, there is a classic example of compacted topsoil. Note how the soil structure in the upper part of the profile has completely collapsed. This limits root growth and exploitation of soil water and nutrients by crops (JJHVDA).

The impact of subsoil compaction is most prominent in years with extreme dry or wet periods. Crop yield reductions of more than 35% have been measured. Subsoil compaction proves to be very persistent, even in subsoils with shrinkage and swelling or annual deep freezing. Reduced crop yields and reduced nitrogen content in crops were detected 17 years after a single compaction event with wheel loads of 50 kN or 5,000 kg.

Scale

All agricultural soils in developed countries display some degree of subsoil compaction. Estimates in 1991 suggest that the area of degradation attributable to soil compaction in Europe may equal or exceed 33 million hectares (ha). Recent research has showed that compaction is the most widespread kind of soil physical soil degradation in central and eastern Europe. About 25 million ha were deemed to be lightly compacted while a further 36 million ha were more severely affected.

Well-structured soils combine good physical soil properties with high strength. Sandy soils with a single grain structure and compacted massive soils can be very strong. However, rootability and soil physical properties are then often bad. Roots have a binding action and increase the elasticity and resistance of a soil to compaction

Soil moisture has a dominant influence on soil compactibility. Dry structured soils are strong with low compactibilty. However, extremely dry sandy soils can be deformed and compacted rather easily. As the moisture content increases, compactibility increases until the moisture content is approximately at the field capacity point, when a condition known as the optimum moisture content for compaction is reached. At still higher moisture content, the soil becomes increasingly incompactible as water fills ever more pore space. Although the compaction of an overloaded wet soil may be minimal, plastic flow may result in the complete destruction of soil structure and macro-pores. Increasing the organic matter content tends to reduce soil compactibility and to increase its elasticity.

Areas degraded by soil compaction are increasing because wheel loads in agriculture are still increasing (Image above). Twenty years ago wheel loads of 50 kN (5000 kg) were considered very high. Nowadays wheel loads of up to 130 kN are used during the harvesting of sugar beet. Modern self-propelled slurry tankers with injection equipment with wheel loads of 90 – 120 kN are used in early spring on wet soils. Large tyres with an inflation pressure of about 200 kPa are needed to carry such high wheel loads. Even on moderate strong soil, compaction of up to 80 cm below the surface have been measured under such loads. The result is that the soil is increasingly compacted to ever-greater depth. The conclusion is that European soil is more threatened than ever (JJHVDA).

Solution

It is almost impossible to avoid topsoil compaction. On the other hand, tillage and natural processes can re-loosen the topsoil. Subsoil compaction is much more persistent and difficult to remove. Artificial loosening of the subsoil has proven to be disappointing. The loosened subsoil is recompacted very easily and many physical properties are strongly reduced.

Subsoil compaction should be prevented instead of being repaired or compensated. Even on weak soils, relatively high wheel loads are possible by using large tyres with low inflation pressures or well-designed tracks. Subsoil compaction during ploughing can be prevented by using improved steering systems and adapted ploughs allowing the tractor to drive with all wheels on the untilled land. It is also possible to concentrate wheel loads on permanent traffic lanes and limit the compaction to these sacrificed wheel ways. By using gantries, the sacrificed area can be limited. However, these solutions are rarely used because of short-term economical constraints, lack of awareness, and negligence because the damage to the subsoil is not readily visible. Also the limited knowledge and data on soil strength under dynamic loading makes prevention of subsoil compaction difficult.

Provisional map of inherent susceptibility of subsoil in Europe to compaction, based on soil properties alone. Further input data are required on climate and land use before vulnerability to compaction of subsoil in Europe can be inferred from the susceptibilities shown here. Some of the very high areas (red) correspond to peat soils that are not subjected to "normal" agricultural practices. However, it is worth including the peat heaths and forests of Europe as they are often used for forestry and can be compacted by heavy timber harvesting machines and off-road vehicles (RJ).

Soil Susceptibility to Compaction

Soil compaction is the rearrangement of soil aggregates and/or particles in a denser way when the voids and pores mainly between the aggregates and particles become smaller or even missing in comparison with the arrangement of similar but not compacted soil. The orientation, size and shape of soil aggregates are evidence of compaction of the soil. Aggregates are arranged with the longer side in a horizontal way (platy structure), they do not have a round shape but one side is much longer than the other and, depending on intensity of compaction, they can be totally destroyed if the compaction is too severe.

Soil susceptibility to compaction is the probability that soil becomes compacted when exposed to compaction risk. It can be low, medium, high and very high depending on soil properties and a set of external factors like climate, soil use, etc. An actualised version of the Map of Natural Susceptibility of Soil to Compaction has been elaborated for evaluation and delineation of priority areas connected to soil compaction at European scale. Soil compaction together with erosion, organic matter decline, salinisation and landslides belongs to the main threats to soil. For these threats the risk area has to be delineated as mentioned in the Thematic Strategy for Soil protection adopted by European Commission on 22 of September, 2006. Knowing priority areas will help politicians as well as soil managers and environmentalists to manage soil use in these areas in such a way that the compaction-threat can be prevented or diminished.

This soil compaction map is exclusively dealing with soil susceptibility to compaction, which is very important for prevention purposes and does not show the real status in European soils. The reasons for this are that there are not enough data for a real status evaluation and significant changes in soil environment during the year when also the real status of soil concerning soil compaction is changing and might be the reason of possible discrepancies in making decisions how to manage touched soils.

For the map at European scale, the soil susceptibility to compaction is more realistic and useful for taking proper sanitary, preventive and management decisions.

Landslides

Principles

A landslide is the gravitational movement of a mass of rock, earth or debris down a slope. Landslides are usually classified on the basis of the material involved (rock, debris, earth, mud) and the type of movement (fall, topple, avalanche, slide, flow, spread). Thus, the generic term landslide also refers to mass movements such as rock falls, mudslides and debris flows. Volcanic mudflows and debris flows are also called lahars.

Shallow landslides usually involve only the soil layer and upper regolith zone, while deep-seated landslides additionally involve bedrock at higher depth. Landslide volume can vary from some tens of cubic metres to several cubic kilometres for giant landslides, while landslide speed may range from a few centimetres per year for slow-moving landslides to tens of kilometres per hour for fast, highly destructive landslides. According to the state of activity or movement, existing landslides can be classified as active, dormant (potentially reactivated) or inactive (often relict or fossil). Landslides are generally induced when the shear stress on the slope material exceeds the material's shear strength. The occurrence and reactivation of landslides is conditioned by a number of terrain and geo-environmental factors related to bedrock and soil properties, weathering conditions, jointing and structure, slope morphology, land cover/use, surface and ground water flow, etc.

Landslides can be triggered by natural physical processes such as heavy or prolonged rainfall, earthquakes, volcanic eruptions, rapid snow melt, slope undercutting by rivers or sea waves and permafrost thawing. They can also be triggered by man-made activities such as slope excavation and loading (e.g. road and buildings construction, open-pit mining and quarrying), land use changes (e.g. deforestation), rapid reservoir drawdown, irrigation, blasting vibrations, water leakage from utilities, etc, or by any combination of natural and/or man-induced processes.

Impacts

Landslides are a major hazard in most mountainous and hilly regions as well as in steep river banks and coastlines. Their impact depends largely on their size and speed, the elements at risk in their path and the vulnerability of these elements. Every year landslides cause fatalities and result in large damage to infrastructure (roads, railways, pipelines, artificial reservoirs, etc.) and property (buildings, agricultural land, etc.).

Large landslides in mountainous areas can result in landslide dams blocking river courses. These natural dams cause valley inundation upstream and can be subsequently breached by lake water pressure, hence generating deadly flash floods or debris flows downstream. Submarine and large coastal cliff landslides can trigger tsunami, as can landslides in lake and reservoir shores.

Landslides can also affect mine waste tips and tailings dams and landfills, causing fatalities and contaminating soils and surface and ground water

In areas affected by landslides, these are a major source of soil erosion and sediment yield to valleys and rivers.

Landslides in Europe

Landslide Occurrence

Landslides occur in many different geological and environmental settings across Europe. For example, large rockfalls, rockslides, rock avalanches and debris flows dominate in the Alps and steep slopes in other mountain ranges; slides and flows abound in flysch belts of Slovakia, Czech Republic, Poland, Italy, Spain, France and other countries; slides of various types are numerous on cliffs and steep slopes in Southern and Eastern England's coast and Bulgaria's Northern Black Sea coast; shallow slides and mudflows occur on Ireland's peat slopes; slides and lateral spreads do as well on gentle slopes in quick clays in Sweden and Norway; flows and slides also typically occur in clay-rich sediments and sedimentary sequences in Tertiary basins as well as on river banks, etc.

Intense and/or long-lasting rainfall represents the most frequent trigger of landslides in continental Europe. However, earthquakes are also responsible for some large landslides. Human activities are also the cause of many slope failures in infrastructure and built-up areas.

Landslides are a major factor of landscape evolution in mountainous and hilly regions in Europe. In addition to causing extensive erosion and sediment yield in these regions, large landslides have been responsible for the creation of many lakes in the Alps and other mountain ranges by damming river valleys.

Examples of this include the lakes of Santa Croce, Antrona (formed in 1642), Alleghe (formed in 1771) and Scanno in Italy, Eibsee and Obersee in Germany, Blindsee in Austria, Vallon in France (formed in 1943), and Sils, Silvaplana, Oeschinen and Davos in Switzerland. Most landslide dams, however, have often formed temporary lakes that have later breached the dam causing catastrophic flash floods and debris flows.

Today, hills in some alpine valley bottoms are remnants of large deposits from giant landslides (e.g. from prehistoric rock avalanches with volumes even in excess of 1 km^3 such as Flims, Sierre and Tamins in Switzerland, Köfels, Fernpass and Tschirgant in Austria, etc.). Many landslide-dammed lakes have been progressively filled with sediments, thus also modifying the valley environment. Unfortunately, the hazard of river damming from landslides still exists in these regions: outstanding examples are those of La Clapière and Séchilliene rockslides in the French Alps, whose potential movement acceleration threatens communities located far from the unstable slopes. On the other hand, landslides in steep coastal areas including cliffs accelerate erosion and subsequent cliff retreat by sea waves.

Giant subaerial landslides are not exclusive of the Alps. They have also occurred in prehistoric times in areas such as southern Crimea in Ukraine, Isle of Skye in UK, and especially in the Canary Islands, Spain. In the latter, a number of huge debris avalanches entered the ocean triggering tsunami. Evidence of large tsunami are also found in Scotland and other coastal areas bordering the Norwegian Sea, mainly attributed to the Storegga submarine megaslide (ca. 3,500 km^3) off the west coast of Norway. In the Mediterranean, landslide-triggered tsunami have been observed dominantly in the Corinth Gulf, Greece, and the Aeolian Islands, Italy. Recent examples include the local tsunami caused by the collapse of the Nice airport embankment in 1979 and the small to moderate tsunami produced by a landslide on the Stromboli Island flank in 2002.

Nowadays, population growth and expansion into landslide-prone areas is raising landslide risk in Europe. In addition, an increase of landslide events is expected in the future due to climate change.

Major Historic Landslide Disasters

There is a long record of landslide disasters in historical and recent times in Europe causing many fatalities and high economic losses. Major disasters include, among others, those of Goldau (1806), Elm (1881) and Gondo (2000) in Switzerland; Piuro (1618), Antronapiana (1642), Roccamontepiano (1765), Monte Antelao (1814, 1925), Vajont (nearly 2000 killed by reservoir wave caused by man-induced landslide in 1963) and, more recently, Valpola (1987), rainfall-triggered multi-landslide events in Piedmont region (1994), Sarno and Quindici (1998) and Messina (2009), and the train accident at Laces (2010), all in Italy; Granier (1248) and Plateau d'Assy (1970) in France; Felanitx (1844), Azagra (1874) and Granada province (earthquake-triggered multi-landslide event, 1884) in Spain; Mount Dobratsch (1348) in Austria; Getå (1918) and Tuve (1977) in Sweden; Verdalen (1893) and landslide-triggered local tsunami at Loen (1905, 1936) and Tafjord (1934) in Norway.

In addition, landslides occurring in mine waste tips and tailings dams have been the origin of the catastrophes of Sgorigrad, Bulgaria (1966), Aberfan (1966) in Wales, UK, and Stava (1985) in Trento, Italy.

Lessons learnt from the management of a number of landslide disasters occurred in Europe in the 1990s and early 2000s are reported here

Landslides and the EU Soil Thematic Strategy

Landslides are one of the soil threats considered in the EU Thematic Strategy for Soil Protection and the related Proposal for a Soil Framework Directive. The Strategy calls for actions and means for the protection and sustainable use of soils as a physical platform on which human activities are developed. The proposed Directive, in turn, will be the Strategy implementing tool. This will mainly require to identify landslide and other soil threat risk areas in the European Union, set risk reduction targets for those areas and establish programmes of measures by Member States to achieve them.

JRC Role and Expertise

The JRC Soil Action provides scientific and technical support to the European Commission Services for the implementation of the EU Thematic Strategy for Soil Protection, both through its own work activities and in collaboration with national research organisations and mapping agencies. Our main activities and expertise include harmonisation of methods for landslide mapping and zoning in Europe (inventory, susceptibility, hazard and risk) at various scales, development of satellite, airborne and ground-based remote sensing techniques for landslide mapping and long term monitoring, analysis of lessons learnt from management of past landslide disasters, and spatial database management. Find more information here

The JRC coordinates a European landslide expert group and participates in the EU Framework Programme 7 project SAFELAND (Living with landslide risk in Europe: Assessment, effects of global change, and risk management strategies). JRC has also participated in previous EU landslide-related research projects including GALAHAD, MUSCL, RUNOUT, ENVASSO and RAMSOIL.

JRC is a member of the International Consortium on Landslides (ICL) and forms part of the Organising Committee of the 2nd World Landslide Forum, to be held on 3-9 October 2011 at FAO headquarters, Rome. JRC staff will convene session L04: Landslide inventory and susceptibility and hazard zoning.

JRC landslide experts are also members of the European Centre on Geomorphological Hazards (CERG), a specialised research network of the Council of Europe's EUR-OPA Major Hazards Agreement co-operation platform.

Chapter 7
Soil Sealing

Soil Sealing is the loss of soil resources due to the covering of land for housing, roads or other construction work.

The covering of the soil surface with impervious materials as a result of urban development and infrastructure construction is known as soil sealing. The term is also used to describe a change in the nature of the soil leading to impermeability (e.g. compaction by agricultural machinery). Sealed areas are lost to uses such as agriculture or forestry while the ecological soil functions are severely impaired or even prevented (e.g. soil working as a buffer and filter system or as a carbon sink). In addition, surrounding soils may be influenced by change in water flow patterns or the fragmentation of habitats. Current studies suggest that soil sealing is nearly irreversible.

The greatest impacts of soil sealing are observable in urban and metropolitan areas. The Figures down illustrate the areas in Europe where the rates of soil sealing are high and where the greatest pressures are likely to occur. In already intensively urbanised countries like Holland or Germany the rate of soil loss due to surface sealing is high. There is little space for further urbanisation. Most of the growth will presumably take place within or on the edge of the suburban areas. In the Mediterranean region, soil sealing is a particular problem along the coasts where rapid urbanisation is associated with the expansion of tourism. Very high rates of sealing are now predicted for countries like Portugal, Finland or Ireland where urbanisation levels are generally low.

In Central and Eastern Europe soil sealing has been comparatively modest in the past decades. An accelerated increase of built-up areas can be recorded as a consequence of the political and economic changes during the late 1980s. Rural populations migrated to the cities and new settlements were developed. Rising pressures on soil can be expected in the course of a strengthened economic growth in these countries. Generally the enlargement of the EU and the integration of new countries in the common market will lead to a heightened movement of people and transport of goods. More infrastructure will be built in order to ensure a good connection between peripheral regions and the centre.

Built-up areas have been mainly enlarged at the expense of agricultural land. Progressive soil sealing will take place especially for Western Europe where the area of built-up land increases at a faster rate than the population. Besides the influence of tourism, the rising demand for land resources can be mainly caused by changes in population behaviour such as people's preference for living outside the city centres, an increased demand for bigger houses or out of town developments such as supermarkets, leisure centres and associated development of transport infrastructure.

Spatial planning strategies determine to a great extent the progression of soil sealing. Unfortunately neither the economical nor the ecological or the social effects of irreplaceable soil losses have been considered adequately so far.

In the meantime the necessity to include environmental concerns and objectives in spatial planning, in order to reduce the effects of uncontrolled urban expansion, is widely recognised in the EU. A rational land-use planning to enable the sustainable management of soil resources and the limiting of sealing of open space is demanded. Possible measures include the redevelopment of brown-fields and the rehabilitation of old buildings.

Soil Contamination

Soil contamination is the occurrence of pollutants in soil above a certain level causing a deterioration or loss of one or more soil functions. Also, Soil Contamination can be considered as the presence of man-made chemicals or other alteration in the natural soil environment. This type of contamination typically arises from the rupture of underground storage tanks, application of pesticides, percolation of contaminated surface water to subsurface strata, leaching of wastes from landfills or direct discharge of industrial wastes to the soil. The most common chemicals involved are petroleum hydrocarbons, solvents, pesticides, lead and other heavy metals. The occurrence of this phenomenon is correlated with the degree of industrialisation and intensity of chemical usage.

Soil Biodiversity

Decline in Soil Biodiversity is the reduction of forms of life living in soils, both in terms of quantity and in variety.

Soil biodiversity is a term used to describe the variety of life below-ground. The concept is conventionally used in a genetic sense and denotes the number of distinct species (richness) and their proportional abundance (evenness) present in a system, but may be extended to encompass phenotypic (expressed), functional, structural or trophic diversity. The total biomass below-ground generally equals or exceeds that above-ground, whilst the biodiversity in the soil always exceeds that on the associated surface by orders of magnitude, particularly at the microbial scale. A handful of grassland soil will typically support tens of thousands of genetically distinct prokaryotes (bacteria, archaea) and hundreds of eukaryotic species across many taxonomic groups. The soil biota plays many fundamental roles in delivering key ecosystem goods and services, and is both directly and indirectly responsible for carrying out many important functions.

Ecosystems goods provided by soil biota:
- food production
- fibre production
- provision of secondary compounds (e.g. pharmaceuticals / agrochemicals).

Ecosystems services provided by soil biota :
- driving nutrient cycling and regulation of water flow and storage
- regulation of soil and sediment movement and regulation of other biota (including pests and diseases)
- detoxification of xenobiotics and pollutants and regulation of atmospheric composition.

The Value of Soil Biodiversity

Soil biodiversity carries a range of values that depend on the perspective from which they are being considered. These include:
- Functional value, relating to the natural services that the soil biota provides, the associated preservation of ecosystem structure and integrity, and ultimately the functioning of the planetary system via connections with the atmosphere and hydrosphere.

- Utilitarian ("direct use") value, which covers the commercial and subsistence benefits of soil organisms to humankind.
- Intrinsic ("non-use") value, which comprises social, aesthetic, cultural and ethical benefits
- Bequest ("serependic") value, relating to future but as yet unknown value of biodiversity to future planetary function or generations of humankind.

The ecological value of soil biodiversity is increasingly appreciated as we understand more about its origins and consequences. The monetary value of ecosystem goods and services provided by soils and their associated terrestrial systems, an entirely human construct which assists putting their significance into an economic context, was estimated in 1997 to be thirteen trillion US dollars ($13 x 1012). The soil biota underwrites much of this value.

In the image, you can view an approximate number and diversity of organisms typically found in a handful of temperate grassland soil (KR & JJIM).

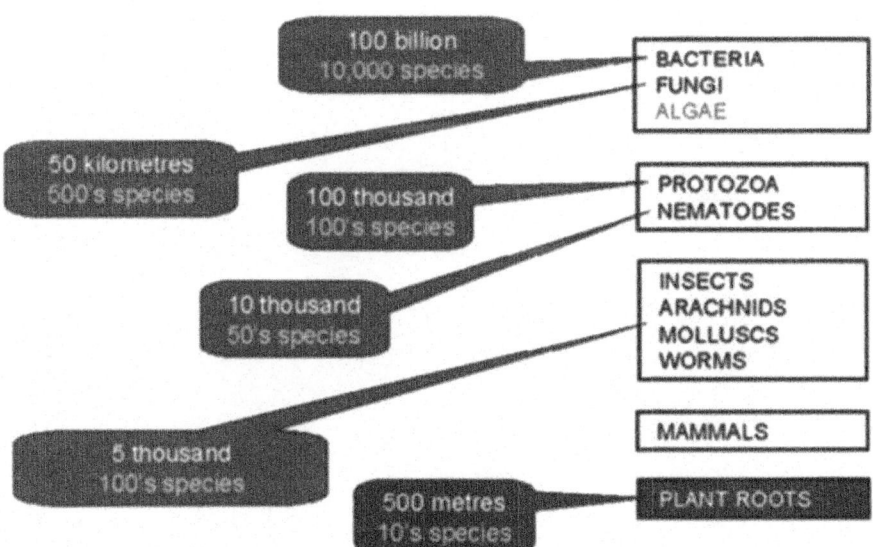

Threats to Soil Biodiversity

A healthy soil biota needs an appropriate habitat. In soil, this is essentially the space denoted by the complex architecture of the pore network, and the associated supply and dynamics of gases, water, solutes and substrates that this framework supports. Hence threats to soil such as erosion, contamination, salinisation and sealing all serve to threaten soil biodiversity by compromising or destroying the habitat of the soil biota.

Management practices that reduce the deposition or persistence of organic matter in soils, or bypass biologically-mediated nutrient cycling also tend to reduce the size and complexity of soil communities. It is however notable that even polluted or severely disturbed soils still support relatively high levels of microbial diversity at least. Specific groups may be more susceptible to certain pollutants or stresses than others, for example nitrogenfixing bacteria that are symbiotic to legumes are particularly sensitive to copper; colonial ants tend not to prevail in frequently-tilled soils due to the repeated disruption of their nests; soil mites are a generally very robust group.

Consequences of Soil Biodiversity

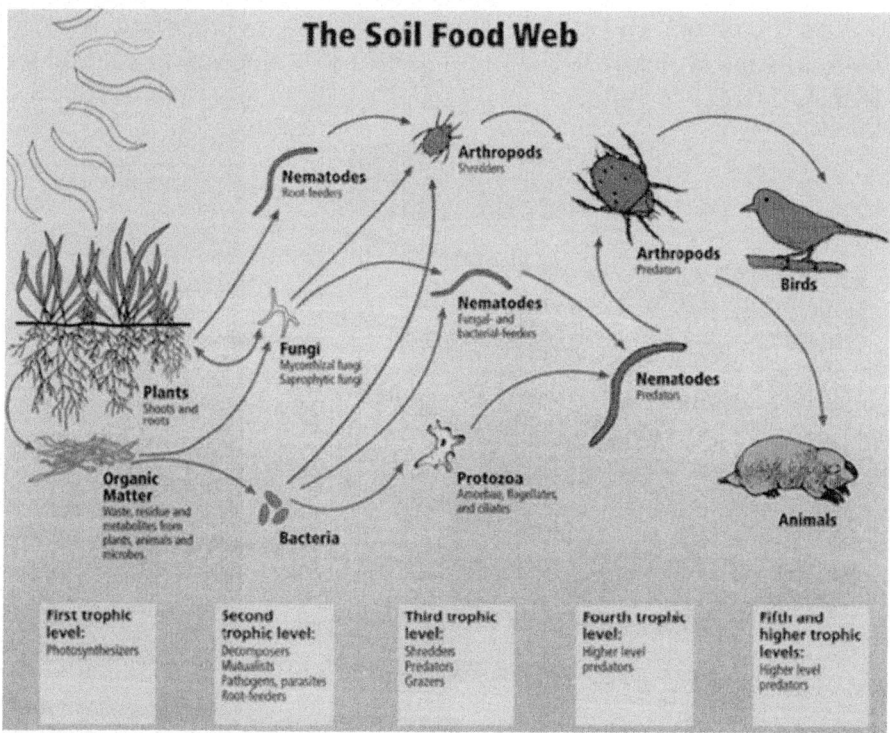

The relationships between biodiversity and function are complex and somewhat poorly understood, even in aboveground situations. The exceptional complexity of belowground communities further confounds our understanding of soil systems. Three important

mechanisms underlying relationships between biodiversity and function are:
- Repertoire: for a biologically-mediated process to occur, organisms that carry out that process must be present; Interactions: most soil organisms have the capacity to directly or indirectly influence other organisms, either positively or negatively;
- Redundancy: the more organisms there are that can carry out a function in a particular soil, the more likely it is that if some are incapacitated or removed the process will remain unaffected; those that remain fill the gap.

There is evidence that soil biotic communities are coupled to their associated vegetation, such that there is a mutual dependence between above-ground and below-ground communities, and hence that compromised soil communities may curtail particular plant assemblages from forming.

In the image, you may view Simplified soil food web. Energy and nutrient elements are transferred from one trophic level to another. Note that there is also a continual movement of material from all trophic levels back to the detritus/organic matter pool and the base of the series (Tugel, A.J. & A.M. Lewandowski, eds., Soil Biology Primer.

Consequences of Decline in Soil Biodiversity

richness per se is of little consequence; rather it is the functional repertoire of the soil biota that is critical. For processes such as decomposition, there is evidence that there is a high degree of redundancy at a microbial level. Other processes, such as nitrification (the oxidation of ammonium), are carried out by a narrower range of bacteria and there is less redundancy in this group, whereas for highly specific symbiotic associations, such as those between orchids and certain mycorrhizal fungi, there is total dependence and hence zero redundancy. A depletion of biodiversity will therefore have differing consequences in relation to different processes.

In some circumstances it has been demonstrated that there are threshold levels of soil diversity below which processes are impaired, although these are usually related to narrow processes and are manifest under experimentally constructed systems of exceptionally low levels of diversity, as opposed to natural systems. From the intrinsic and bequest perspective, any loss of biodiversity is undesirable. Given our limited state of understanding of the consequences of soil biodiversity, it is common sense that a strong precautionary principle needs to be applied.

Soil Sampling

Abundance and quality of the organic carbon (C) are two among the parameters assisting to distinguish soils from surface rocks and loose deposits. Soil scientists widely recognise soil C to be one of the major product of the soil formation, which in turn drives principle soil-ecological functions, e.g., regulatory role of soils in global biogeochemical cycle of C. Soil is the second large organic carbon (C) pool in the Earth (IPCC, 2001).

This pool comprises the long-term C storage measured by 1×10^2-10^3 years. In comparison, the lifetime period of C in vegetation varies within 1×10^2 years and decay of litter takes up to 10 years. Heteroptophic soil respiration exchanges 10% of carbon dioxide (CO^2) of the atmosphere annually (Raich et al., 1995). Therefore the significance of the soil in the global C cycle is difficult to estimate.

The role of soils in the Kyoto protocol is limited only to agriculture, e.g., the protocol underlines the necessity to account the changes in greenhouse gas emissions by sources and removals by sinks in the agricultural soils (Article 3.4, INFCCC, 1998). The protocol is silent regarding possible affect of land use, land use change and forestry on soils (Article 3.3). This missing was furnished later by two supplementary reports: (1) Land Use, Land-Use Change, and Forestry (LULUCF) (IPCC, 2000) and (2) Good Practice Guidance for LULUCF (IPCC, 2003) identifying soil organic C to be an obligatory to account when implementing Articles 3.3 (afforestation, reforestation and deforestation since 1990). However, the general norms provided by these documents on accounting as well as on countrywide reporting are insufficient to be applied in the agricultural fields and forests exclusively. There is a need to develop a reliable method on soil sampling to certify the changes of carbon stock in soils, lack of which might be a serious obstacle for the Kyoto implementation in EU.

The IPCC (2003) proposes to assess the changes of C stock in soils as a difference between two independently estimated spatially-averaged C contents of mineral soils for the reference (baseline) and new (current) observation. The uncertainty of the estimate arrives from the variability of soil parameters, which demand considerable amount of samples to reach required (95%) confidence level of the C stock changes detection. This makes soils involvement in C account expensive and impractical.

The suggested method of soil sampling is different. It introduces area frame sampling combining traditional in agrochemistry composite sampling with random geopositioning of the sampling sites in the field. This method insures a reproducibility of the sampling sites in the followed up samplings, which allows to minimise the amount of the samples to the practical level.

In addition, this method substantialises difference in soils by LULUCF categories, e.g., IPCC oversimplifies representation of soils by viewing only one soil layer (0-30 cm) for cropland, pastures and forests. The change of C stock is proposed to be certified by the weight of C and standard error.

The overall purpose of the present study is to provide simple, acceptable and adaptable guidelines to make soil sampling relevant to the purposes of the Kyoto, namely remain practically feasible accounting soil diversity, statistically sound and economically viable.

Organic Carbon Stock in Mineral Soils of European Union

The "Soil Sampling Protocol to Certify the Changes of Organic Carbon Stock in Mineral Soils of European Union" is complied by a group of scientists within Integrated Sink Enhancement Assessment project and the action MOSES - Monitoring the State of European Soils (Land Management Unit, Institute for Environment and Sustainability, Joint Research Centre of the European Commission) to support implementation of the Kyoto Protocol in EU.

Soil Organic Matter, Green Manures and Cover Crops for Nematode Management

Much of the soil in Florida is very sandy and has very little organic matter. Vegetables, flowers and landscape plants grown in soil that is high in organic matter often are less damaged by nematodes than plants grown in soil with less organic matter content. Any kind of organic soil amendment, including compost, green manures, and lightly incorporated organic mulches, can have this effect. Organic amendments can both improve tolerance of the plant to nematodes and also reduce nematode populations. However, they can not magically eliminate a severe nematode infestation overnight. They are better suited to keeping nematode populations relatively low than reducing high ones. There are several ways in which organic soil amendments may help reduce nematode injury to plants.

Soil Organic Matter

Soil organic matter is any material in the soil that was originally produced by living organisms. At any given time, it consists of a range of materials varying from intact plant material to the substantially decomposed organic materials known as humus. Plant tissues contain a wide range of organic compounds which typically decompose at different rates.

Adding fresh plant material to a soil which has little or no readily decomposable materials immediately starts rapid multiplication of bacteria, fungi, and other microorganisms, which actively decompose the fresh tissue. As the most readily available energy sources (carbohydrates, fats, proteins) in the fresh material are used up, those microorganisms again become relatively inactive.

Most plant-parasitic nematodes feed on and damage plant roots. Many of the negative affects from nematodes are due to the inability of nematode-damaged roots to obtain water and nutrients from soil. Thoroughly decomposed organic matter (called Humus) improves soil texture and the ability of soil to retain water and many plant nutrients. Because humus improves the ability of soil to hold onto water and nutrients it gives plants more opportunity to obtain them. This is how organic amendments improve plant tolerance to nematodes. Additionally, humus promotes better soil structure by causing soil to stick together, and it provides some nutrients as it is slowly decayed by microbial activity. Mulches of organic materials which help keep roots cool and reduce evaporation from soil also help reduce stress, and add humus as they decompose.

Adding Organic Matter to Soil

There are several ways to add organic matter to soil including:
- compost — organic material that has been substantially decomposed under somewhat controlled conditions, so that most of the readily-exploited nutrients are exhausted
- non-composted plant and animal material — common examples are leaves and garden waste, or fresh animal manure
- green manure — fresh plant material grown purposely to be incorporated into the soil on the site where it was grown.

Compost

Compost provides many of the nutrients plants need, and be converted into humus. Since the readily-available nutrients are mostly consumed in the composting process, compost will not stimulate microbial activity to the same degree as fresh plant tissue.

Fresh Organic Matter

Whole plants, green grass clippings and leaves, or fresh animal manures can add organic matter to soil. Material so added can help maintain or raise the organic matter content of the soil, and thus its ability to produce crops, in several ways.

Whole, but not Fresh, Organic Matter

Dry plant materials like wood chips, saw dust, peanut hulls, or pine straw typically need moisture to be added for decomposition to proceed. Because nutrients like nitrogen are typically lacking in these materials, they break down much more slowly than fresh organic matter. Also, soil nutrients may be tied up by microbes attempting to break down these types of materials. Therefore, mixing dry organic materials with the fresh organic matter defined above and/or composting prior to use is generally recommended before using them as a soil amendment.

Green Manure

Green manuring (discussed in detail below) is turning under of a green crop to better the condition of the soil. Material so added, if soil is in proper condition and well managed, can help maintain or raise the organic matter content of the soil, and thus its ability to produce crops, in several ways.

Soil Biological Activity

Decomposing organic matter is food for many soil microbes (fungi, actinomycetes, and bacteria). Some of those creatures are natural enemies of plant nematodes; increasing their numbers enhances "natural" nematode control.

Chemical By-Products

Decomposition of some organic amendments, including some green manures oily plant residues such as cottonseed meal, or meal from certain types of mustard have been shown to release chemicals which are directly toxic to nematodes. Those chemicals may reduce nematode numbers directly, in addition to the other benefits derived from soil organic amendments.

"Nematicidal" Miracle Products

Claims of nematode control are made for many products sold as soil amendments and additives, but objective research data rarely accompany those claims. Research conducted at University of Florida has found most of these do not suppress nematodes any more than the much cheaper organic amendments discussed previously. If a product's claims cannot be supported with evidence from well-designed research, preferably conducted by scientists who are in no way connected with the product, be very cautious about depending on it for nematode control.

Green Manures

Green manuring is the practice of growing plants on the site into which you wish to incorporate organic matter, then turning (tilling, plowing, spading) it into the soil while it is still fresh. The plant material used in this way is called a green manure.

The plant material may or may not be cut free from its roots before being incorporated into the soil, depending on what is needed to be able to handle it. Green manuring is popular among organic gardeners and farmers; it is also used by some conventional growers who recognise its benefits. It is widely adopted simply because adding large amounts of green plant material benefits many soil characteristics, including drainage and water retention, nutrient content and form of storage, and level of microbiological activity in the soil.

- Green manuring supplies soil organic matter, as already discussed.
- It can conserve or even add nutrients. Nitrogen used by the green manure crop has been protected from leaching loss and microbial degradation that could take place if the land were left fallow. Other nutrients such as potassium, phosphorus, magnesium, and iron can be similarly conserved through green manuring. Using a legume can increase soil nitrogen levels through N fixation by the Rhizobium bacteria associated with most legume roots.
- Microbiological benefits—there is a substantial increase in activity of soil fungi and bacteria that do many useful things.

Generally, the more diverse and greater the soil microbial population, the more productive the soil will be.
- Conservation of topsoil against loss by erosion can be a substantial nutrient conservation benefit of green manuring.

Green manuring has also been demonstrated to reduce levels of nematodes in the soil in some cases, in addition to reducing the amount of damage due to nematodes feeding on plant roots.

To be useful as a green manure, a plant should: 1) grow rapidly; 2) produce abundant and succulent tops; 3) grow well in the conditions of the site. The higher the moisture content of the material, the more rapidly will it break down and its benefits be realised. When other conditions are equal, it is better to use a legume because of the nitrogen fixation and the microbial activity it promotes.

Crops that are selected for use as green manures are usually chosen for their ability to grow very rapidly and produce a large mass of top growth at the season and site. Cool season grasses are often used in Florida for fall and winter green manure crops; legumes are often desirable because they naturally convert nitrogen from the air into plant nutrients, thus increasing the levels of that most critical nutrient in the soil when the plant tissues are incorporated into the soil. It is generally best to turn green crops under when their succulence is near the maximum, yet when enough top growth has been produced. Usually this is before flowering.

If a "nematicidal" crop such as those discussed below can be used as a green manure, it may have an even greater effect on nematodes in the soil, by beginning their reduction as a cover crop before it is turned into the soil. For instance, hairy indigo is valuable in both ways in North and Central Florida.

Risks in Green Manuring

Take care to avoid planting green manure crops that can encourage reproduction of nematodes, especially root-knot nematodes. Increasing a nematode population on a green manure crop can offset all benefits that would otherwise be gained. If at all possible, choose crops for green manuring that are known to inhibit root-knot nematodes.

"Nematicidal" Cover Crops

Several plants have been found to help control some kinds of nematodes when grown for several months in soil where they are present, if no good hosts of those nematode species are present. These plants may reduce nematode populations significantly more than fallowing or growing a crop on which the nematodes do not feed. Control sometimes equals that from chemical treatments. In other words, using these plants as cover crops may significantly reduce nematode injury to susceptible crops grown in the next season. They rarely, if ever, protect plants from nematodes when planted among or beside them (companion planting or inter-planting) under field conditions in Florida.

Manure Management and Effects of Manure on the Environment

Manure Management Systems

Manure management systems are highly diverse, among which the following can be distinguished:

-Grazing. Substantial losses through leaching may occur due to the uneven distribution of faeces and urine (urine patches may have a N load equal to 200-550 kg/ha; Van der Meer and Meeuwissen, 1989; Romney et al., 1994). Volatilisation of N may also be considerable (10-25%), but less important since part is deposited on nearby areas, though some of it on non-agricultural land.

Kraals: These enclosures are often used as in-situ fertilization of arable land by moving the kraal regularly. Soil fertility of a larger area, used for grazing, is partially concentrated on the arable land, thus enabling crop production in resource-poor situations. Losses through leaching will be slightly higher than during grazing as equivalent N and K fertilization rates are increased.

Dry lot storage: If urine is not collected and bedding is sparsely used, losses of N and K in particular will be high as most urine is lost. Depending on the storage facilities and storage time of the faeces part of the nutrients in faeces will also be lost through leaching and surface runoff, in the case of a precipitation surplus and uncovered manure heaps. Urine collection will minimise K losses but N losses will often remain high as volatilisation will increase, though this is dependent on climatic conditions, storage time and storage method. Using bedding, with sufficient absorption capacity to capture urine, might reduce N losses with ca. 15% of the mineral N.

Slurry storage: This system of manure storage, where faeces and urine are stored together, is the main system in intensive livestock systems in OECD countries, except for broilers. Volatilisation losses are dependent on the level of ventilation, depth of storage tanks and storage time, but often range between 5 and 35% of the total N excreted.

Lagoons: Lagoon systems are quite common at large livestock farms in Eastern European countries and, to a lesser extent, in Asia, while their importance is growing in the USA. Liquid manure, either before on after separating part of the solids, is treated in an aerobic lagoons. Organic material is decomposed, thereby mineralising part of the nutrients. The liquid phase is either discharged into surface water or used for irrigation. The main problems are related to the discharge into surface water , leaching through the lagoon bottom, and odour. High NH_3 emission will occur as a major part of the N in mineral form, while also high CH_4 and N_2O emissions are also common.

Plastering material for house construction: This is particularly important in Africa, however the amount of manure involved on a global scale is considered to be too insignificant to be discussed here. In this system all nutrients are lost for agriculture.

Fuel: In many developing countries, and particularly in India, manure is an appreciable fuel. If burnt directly, most of the C, and all the N and S will be lost; other nutrients may be recycled to arable land via the application of the ash. The production of biogas from manure is another method to valorize the energetic value of manure. The high water content of the slurry makes it more difficult to handle, and N losses via volatilisation may be high, because most N in slurry is in mineral form. Though strongly promoted (e.g. in China) and applied to some extent in Asia, its present application is still limited mainly due to high investment costs (both for the digester and adjoining equipment) and technical problems (Henglian, 1994).

Feed: Manure could be recycled by feeding it to animals, both livestock and fish (Müller, 1980), but this practice is limited. In addition to widespread reluctance to use manure as feed, probably originating from fear of health hazards, this can be explained by the low nutritive value of most types of manure, except for poultry manure as ruminant feed which is of a reasonable quality (ca. 55-60% TDN, 20-30% CP). Consequently, in intensive production systems where collectable manure is abundant, more economic feed is available, while in production systems where the use of low quality feeds is common, high collection costs and/or opportunity costs (manure as fertilizer or fuel) are prohibiting the use of manure as feed. Therefore, no more attention will be paid to this subject in this report. A recent overview on this subject, however, is given by Sanchez (1994).

Emissions before Manure Application

Stables and manure storage are major sources of ammonia (NH_3) emission. About half of the N in manure (i.e. liquid manure or slurry) is NH_3-N in solution. Because of the high vapour pressure of the NH_3, it will readily volatilise upon exposure of the manure to the air. The greater the exposure, i.e. a larger specific area in contact with the air, the more NH_3 volatilisation.

If manure is stored in direct contact with the soil, the liquid can seep into the soil and into the ground water. The nutrients N, P, K, organic and other compounds can leach into the ground water and the manure storage thus acts as a major source of ground water pollution.

Surface runoff from farmyards, kraals, bomas and manure storage can be an important source of pollution.

Effects on Soil Quality

Manure application to agricultural land involves the addition of all the components of the manure to the soil. An appropriate balance should be maintained between agronomic requirements and negative environmental impacts.

Negative impacts, that could be defined as soil pollution, have to do with the addition of heavy metals, organo-chlorines and too many salts. Also, weed seeds could be spread through manuring the land. On the other hand, manuring almost always has a positive influence on the build up of soil organic matter and thus improves the "intrinsic" fertility of the soil, as well as the soil structure.

After application of manure, decomposition by microorganisms of the organic material will start into carbon dioxide (CO_2), water (H_2O) and minerals of plant nutrients such as N, P, S and metals. The transformation of organically bound elements into plant available nutrients during microbiological decomposition is called mineralisation. Organic matter that remains one year after application is assumed to be part of the soil organic matter and will decompose gradually over the years, releasing plant nutrients in a way that resembles a slow release fertilizer. A more fertile soil, consequently, has less need for mineral fertilizers. The fertilizer industry uses non-renewable resources such as fossil fuel, phosphorus and potassium deposits, and is a source of emissions. Manuring has, in an indirect way, a positive effect on the environment. A small fraction of the added organic material is transformed into organic matter that is resistant to microbiological breakdown, the so-called humus or stable organic matter. Humus contributes to soil fertility by retaining plant nutrients through adsorption. It also acts as binding material in the soil, improving soil structure and is responsible for making clay soil less susceptible to compaction caused by heavy traffic, or a silty soil less susceptible to erosion. In addition, humus increases the water holding capacity and the cation exchange capacity (CEC) of any type of soil.

The heavy metals Cu and Zn have been mentioned as major contaminants from the heavy application of pig slurry (e.g. in part of The Netherlands). Repeated application of large doses of pig slurry to the same plot may lead to Cu and Zn levels in the soil that are toxic, for instance, to soil fauna and sheep. Since the 1978 EC legislation Cu additions to pig feed have been reduced increasingly, to a level of 35 mg per kg for growing and finishing pig feeds. At current levels, Cu and Zn are considered not problematic if P fertilization is in balance with the crop requirements.

There is a danger of incomplete degradation of organo-chlorines by microorganisms. Through manuring, they could be taken up by crops and pose a threat to humans by accumulating somewhere in the food chain (L'Hermite et al., 1993). Many countries have replaced organo-chlorines with organo-phosphates, but residues from insecticides still continue to be the main source of organo-chlorines in feed. Organo-chlorines originating from substances used against ecto-parasites can also be found.

The passage of weed seeds through the digestive tract of animals reduces their germination capability. Some weed species, however, survive. In a stack of farmyard manure, the temperature rises above 55 °C because of the microbiological decomposition of the organic matter and kills weed seeds within three weeks. The germination capability of weed seeds stored in slurry is destroyed only after a period of five months (L'Hermite et al., 1993).

Manure contains much dissolved potassium chloride (KCl) and sodium chloride (NaCl). Repeated application of large amounts of manure in arid or semi-arid climates may easily lead to salinisation of the soil, making it unsuitable for many crops (Sequi and Voorburg, 1993).

Effects on Ground Water and Surface Water Quality

The main dangers of the application of manure are runoff of manure or manure components into surface water and leaching of nitrate (NO_3) and P into the ground water. Mineral N in manure is largely present as NH_3. If, upon application of the manure, it does not volatilise, it will be quickly nitrified, i.e. transformed through microbial action into NO_3. Also, N mineralised from the organic fraction of the manure, will readily be nitrified. As NO_3 is an anion that is not adsorbed by clay minerals or soil organic matter, it is easily leached in case of a precipitation surplus. This holds good for NO_3-N from manure, and for that originating from mineral fertilizers or from decomposed soil organic matter. If ground water concentrations of NO_3 become too high, it is unsuitable for drinking water. Under certain conditions, ground water can flow into surface water. In brackish and salt water in particular high NO_3 concentrations in surface water will lead to eutrophication. Under certain conditions this may lead to excessive growth of algae, causing oxygen shortage and consequently the death of fish.

Phosphorus is not nearly as mobile in the soil as NO_3 and therefore much less susceptible to leaching. Nevertheless, leaching of P can occur under certain conditions. Many sandy soils in The Netherlands have become "saturated" with phosphate (P_2O_5) after many years of heavy doses of manure. When saturated, the soil loses part of its capacity for P retention and leaching occurs. If P flows into the ground water and subsequently into surface water, the same problems as described above for NO_3 will occur. Note, P causes eutrophication in freshwater bodies in particular.

Effects on Air Quality

Two processes involving N and one involving carbon (C) from manuring have an important effect on air quality. First, surface application of manure, particularly liquid manure, may cause substantial losses of NH_3 by volatilisation. In the Netherlands, for instance NH_3 volatilisation from manure is a major contributor to acid deposition. Unlike SO_2, a contributor to acid deposition from cars and industry, most of the emitted NH_3 is deposited near the emission source. Forests near regions with a high livestock density, are in a poor condition because of soil acidification caused by NH_3 deposition originating from the livestock industry. Acidification may lead to mobilisation of aluminium (Al) ions, which are very toxic to fish, disturbs the nutrient uptake of plants and trees, and enhances sensitivity to stress factors like drought and fungi. Besides the acidifying effect, NH_3 deposition accounts for a considerable N load to the environment, causing eutrophication problems and N enrichment of the soils in nature reserves. The last mentioned can cause undesirable changes in species composition (important for biodiversity).

Second, denitrification of NO_3 by microorganisms is possible under an aerobic conditions when N_2 is formed, but giving off a by-product N_2O, a gas that affects the ozone layer. Although quantitative data are scarce, animal excreta and arable land may be important sources of N_2O globally.

A third important air pollutant is methane (CH_4), formed upon decomposition of manure under an aerobic conditions. If stored manure is disturbed, CH_4 will escape into the atmosphere and eventually, like N_2O, affect the ozone layer.

In addition to CH_4 formation in manure storage, the use of manure in flooded rice production (an aerobic conditions) and CH_4 formation in the rumen of ruminants are important sources of CH_4 emission. Methane and its consequences are discussed in detail in one of the other reports.

Odour has a negative effect on the air quality, affecting animals and people in closed stables as well as people near farms producing or applying manure. Especially in combination with dust odour may cause serious health problems. In the USA it is estimated that 70% of the workers in closed animal stables suffer from respiratory illness (Scialabba, 1994), this is one of the main causes of labour disability of pig farmers in the Netherlands. Hydrogen sulphide and ammonia are the main contributors to odour problems, though a dozen other compounds, like fatty or organic acids, phenols, etc. may also play a role. Most of the odour comes from the an aerobic decomposition of manure. This aspect of manure is not further discussed in this report.

Effects on Crops

Manure is applied to agricultural land chiefly because of its fertilizing value. Animal manure supplies all major nutrients (N, P, K, Ca, Mg, S,) necessary for plant growth, as well as micronutrients (trace elements), hence it acts as a mixed fertilizer . The fertilizing effect on crops can be compared to the effect of mineral fertilizers, and expressed in working coefficients. If, for example, the N in pig slurry on maize is half as effective in terms of yield increase as the N from ammonium nitrate (which is the reference chemical fertilizer), the working coefficient is 0.5.

Manure application in a given year will influence not only crops grown that year, but also crops in subsequent years, because decomposition of the organic matter is not completed within one year. Working coefficients for subsequent years could be determined as well. Therefore, the application of manure, thus, saves mineral fertilizers for various nutrients. This illustrates that nutrients from animal manure can be substituted for mineral fertilizers and which is far better for the environment.

A disadvantageous aspect of the uptake of components from manure by the crop is over-dosage, which can lead to the absorption by plants of non-degradable components such as heavy metals (Cu, Zn) and organo-chlorines. These components can accumulate in the food chain and become a health hazard.

Choosing a Soil Amendment

Quick Facts:
- Soil amendments improve the physical properties of soils.
- Amendments are mixed into the soil. Mulches are placed on the soil surface.
- The best soil amendments increase water- and nutrient-holding capacity and improve aeration and water infiltration.
- Wood products can tie up nitrogen in the soil.
- Sphagnum peat is superior to Colorado mountain peat.
- When using biosolids, choose Grade 1 biosolids.

A soil amendment is any material added to a soil to improve its physical properties, such as water retention, permeability, water infiltration, drainage, aeration and structure. The goal is to provide a better environment for roots.

To do its work, an amendment must be thoroughly mixed into the soil. If it is merely buried, its effectiveness is reduced, and it will interfere with water and air movement and root growth.

Amending a soil is not the same thing as mulching, although many mulches also are used as amendments. A mulch is left on the soil surface. Its purpose is to reduce evaporation and runoff, inhibit weed growth, and create an attractive appearance. Mulches also moderate soil temperature, helping to warm soils in the spring and cool them in the summer. Mulches may be incorporated into the soil as amendments after they have decomposed to the point that they no longer serve their purpose.

Organic vs. Inorganic Amendments

There are two broad categories of soil amendments: organic and inorganic. Organic amendments come from something that is or was alive. Inorganic amendments, on the other hand, are either mined or man-made. Organic amendments include sphagnum peat, wood chips, grass clippings, straw, compost, manure, biosolids, sawdust and wood ash. Inorganic amendments include vermiculite, perlite, tire chunks, pea gravel and sand.

Not all of the above are recommended by Colorado State University. These are merely examples. Wood ash, an organic amendment, is high in both pH and salt. It can magnify common Colorado soil problems and should not be used as a soil amendment. Don't add sand to clay soil — this creates a soil structure similar to concrete.

Organic amendments increase soil organic matter content and offer many benefits. Organic matter improves soil aeration, water infiltration, and both water- and nutrient-holding capacity. Many organic amendments contain plant nutrients and act as organic fertilizers. Organic matter also is an important energy source for bacteria, fungi and earthworms that live in the soil.

Application Rates

If your soil has less than 3 percent organic matter, then apply 3 cubic yards of your chosen organic amendment per 1,000 square feet. To avoid salt buildup, do not apply more than this. Retest your soil before deciding whether to add more soil amendment.

Wood Products

Wood products can tie up nitrogen in the soil and cause nitrogen deficiency in plants. Microorganisms in the soil use nitrogen to break down the wood. Within a few months, the nitrogen is released and again becomes available to plants. This hazard is greatest with sawdust, because it has a greater surface area than wood chips. If you plan to apply wood chips or sawdust, you may need to apply nitrogen.

If you plan to apply wood chips or sawdust, you may need to apply nitrogen fertilizer at the same time to avoid nitrogen deficiency.

Sphagnum Peat vs. Mountain Peat

Sphagnum peat is an excellent soil amendment, especially for sandy soils, which will retain more water after sphagnum peat application. Sphagnum peat is generally acid (i.e., low pH) and can help Gardeners grow plants that require a more acidic soil. Colorado mountain peat is not as good a soil amendment. It often is too fine in texture and generally has a higher pH.

Mountain peat is mined from high-altitude wetlands that will take hundreds of years to rejuvenate, if ever. This mining is extremely disruptive to hydrologic cycles and mountain ecosystems. Sphagnum peat is harvested from bogs in Canada and the northern United States. The bogs can be revegetated after harvest and grow back relatively quickly in this moist environment.

Are Biosolids Safe?

Biosolids are byproducts of sewage treatment. They may be found alone or composted with leaves or other organic materials. The primary concerns about biosolids are heavy metal content, pathogen levels and salts. To avoid excessive levels of heavy metals and to ensure that pathogens have been killed, always choose a Grade 1 biosolid. While Grade 1 biosolids are acceptable for food Gardens, do not use them on root Crops because they will come in direct contact with the edible portion of the plant. Do not use biosolids below Grade 1.

Manure vs. Compost

Fresh manure can harm plants due to elevated ammonia levels. To avoid this problem, use only aged manure (at least six months old). Pathogens are another potential problem with fresh manure, especially on vegetable Gardens. Compost manure for at least two heating cycles at 130 to 140 degrees F to kill any pathogens before applying the manure to vegetable Gardens. Most home composting systems do not sustain temperatures at this level. Home-composted products containing manure are best used in flower Gardens, shrub borders and other nonfood Gardens.

During composting, ammonia gas is lost from the manure. Therefore, nitrogen levels may be lower in composted manure than in raw manure. On the other hand, the phosphorus and potassium concentrations will be higher in composted manure. Modify fertilizer practices accordingly. Salt levels also will be higher in composted manure than in raw manure. If salt levels are already high in your

Garden soil, do not apply manures. Other composts are available that are made primarily from leaf or wood products alone or in combination with manures or biosolids.

Factors to Consider When Choosing an Amendment

There are at least four factors to consider in selecting a soil amendment:
- how long the amendment will last in the soil,
- soil texture,
- soil salinity and plant sensitivities to salts, and
- salt content and pH of the amendment.

Laboratory tests can determine the salt content, pH and organic matter of organic amendments. The quality of bulk organic amendments for large-scale landscape uses can then be determined.

Longevity of the Amendment

The amendment you choose depends on your goals.
- Are you trying to improve soil physical properties quickly? Choose an amendment that decomposes rapidly.
- Do you want a long-lasting improvement to your soil? Choose an amendment that decomposes slowly.
- Do you want a quick improvement that lasts a long time? Choose a combination of amendments.

Table 1: Decomposition rate of various amendments.

Amendment	*Decomposition rate*
Grass clippings, manures	Rapid decomposition (days to weeks)
Composts	Moderate decomposition (about six months)
Wood chips (redwood, cedar), hardwood bark, peat	Slow decomposition (possibly years)

Soil Texture

Soil texture, or the way a soil feels, reflects the size of the soil particles. Sandy soils have large soil particles and feel gritty. Clay soils have small soil particles and feel sticky. Both sandy soils and clay soils are a challenge for Gardeners. Loam soils have the ideal mixture of different size soil particles. When amending sandy soils, the goal is to increase the soil's ability to hold moisture and store nutrients. To achieve this, use organic amendments that are well decomposed, like composts or aged manures. With clay soils, the goal is to improve soil aggregation, increase porosity and permeability, and improve aeration and drainage. Fibrous amendments like peat, wood chips, tree bark or straw are most effective in this situation.

More specific recommendations. Because sandy soils have low water retention, choose an amendment with high water retention, like peat, compost or vermiculite. Clay soils have low permeability, so choose an amendment with high permeability, like wood chips, hardwood bark or perlite. Vermiculite is not a good choice for clay soils because of its high water retention.

Table 2: Permeability and water retention of various soil types.

Soil Texture	Permeability	Water Retention
Sand	high	low
Loam	medium	medium
Silt	low	high
Clay	low	high

Table 3: Permeability and water retention of various soil amendments.

Amendment	Permeability	Water Retention
Fibrous		
Peat	low-medium	very high
Wood chips	high	medium low
Hardwood bark	high	medium low
Humus		
Compost	low-medium	medium-high
Aged manure	low-medium	medium
Inorganic		
Vermiculite	high	high
Perlite	high	low

Soil Salinity and Plant Sensitivity to Salts

Some forms of compost and manures can be high in salts. Avoid these amendments in soils that are already high in salts (above 3 mmhos/cm) or when growing plants that are sensitive to salts. Raspberry, strawberry, bean, carrot, onion, Kentucky bluegrass, maple, pine, viburnum and many other landscape plants are salt sensitive. In such cases, choose sphagnum peat or ground leaves instead of compost or manures.

Salt Content and pH of the Amendment

Always beware of salts in soil amendments. High salt content and high pH are common problems in Colorado soils. Therefore, avoid amendments that are high in salts or that have a high pH. Amendments high in salts and/or pH include wood ash, Colorado mountain peat and composted manures. An amendment with up to 10 mmhos/cm total salts is acceptable if well mixed into low-salt soils (less than 1 mmhos/cm). Amendments with a salt content greater than 10 mmhos/cm are questionable. Choose a low-salt amendment for soils testing high in salts. Sphagnum peat and compost made from purely plant sources are low in salts and are good choices for amending Colorado soils. Ask for an analysis of the organic amendments that you are considering, and choose your amendments wisely. If no analysis is available, test a small amount of the amendment before purchasing a large quantity.

Organic Farming : Manures

Manures

Manures are plant and animal wastes that are used as sources of plant nutrients. They release nutrients after their decomposition. The art of collecting and using wastes from animal, human and vegetable sources for improving crop productivity is as old as agriculture. Manures are the organic materials derived from animal, human and plant residues which contain plant nutrients in complex organic forms. Naturally occurring or synthetic chemicals containing plant nutrients are called fertilizers. Manures with low nutrient, content per unit quantity have longer residual effect besides improving soil physical properties compared to fertilizer with high nutrient content. Major sources of manures are:

1. Cattle shed wastes-dung, urine and slurry from biogas plants
2. Human habitation wastes-night soil, human urine, town refuse, sewage, sludge and sullage
3. Poultry Jitter, droppings of sheep and goat
4. Slaughterhouse wastes-bone meal, meat meal, blood meal, horn and hoof meal, Fish wastes
5. Byproducts of agro industries-oil cakes, bagasse and press mud, fruit and vegetable processing wastes etc
6. Crop wastes-sugarcane trash, stubbles and other related material

7. Water hyacinth, weeds and tank silt, and
8. Green manure crops and green leaf manuring material

Manures can also be grouped, into bulky organic manures and concentrated organic manures based on concentration of the nutrients.

Bulky Organic Manures

Bulky organic manures contain small percentage of nutrients and they are applied in large quantities. Farmyard manure (FYM), compost and green-manure are the most important and widely used bulky organic manures. Use of bulky organic manures has several advantages:

(1) They supply plant nutrients including micronutrients
(2) They improve soil physical properties like structure, water holding capacity etc.,
(3) They increase the availability of nutrients
(4) Carbon dioxide released during decomposition acts as a CO_2 fertilizer and
(5) Plant parasitic nematodes and fungi are controlled to some extent by altering the balance of microorganisms in the soil.

Farmyard Manure

Farmyard manure refers to the decomposed mixture of dung and urine of farm animals along with litter and left over material from roughages or fodder fed to the cattle. On an average well decomposed farmyard manure contains 0.5 per cent N, 0.2 per cent P_2O_5 and 0.5 per cent K_2O. The present method of preparing farmyard manure by the farmers is defective. Urine, which is wasted, contains one per cent nitrogen and 1.35 per cent potassium. Nitrogen present in urine is mostly in the form of urea which is subjected to volatilisation losses. Even during storage, nutrients are lost due to leaching and volatilisation. However, it is practically impossible to avoid losses altogether, but can be reduced by following improved method of preparation of farmyard manure. Trenches of size 6 m to 7.5 m length, 1.5 m to 2.0 m width and 1.0 m deep are dug.

All available litter and refuse is mixed with soil and spread in the shed so as to absorb urine. The next morning, urine soaked refuse along with dung is collected and placed in the trench. A section of the trench from one end should be taken up for filling with daily collection. When the section is filled up to a height of 45 cm to 60 cm above the ground level, the top of the heap is made into a dome and plastered with cow dung earth slurry. The process is continued and when the first trench is completely filled, second trench is prepared.

The manure becomes ready for use in about four to five months after plastering. If urine is not collected in the bedding, it can be collected along with washings of the cattle shed in a cemented pit from which it is later added to the farmyard manure pit. Chemical preservatives can also be used to reduce losses and enrich farmyard manure. The commonly used chemicals are gypsum and superphosphate. Gypsum is spread in the cattle shed which absorbs urine and prevents volatilisation loss of urea present in the urine and also adds calcium and sulphur. Superphosphate also acts similarly in reducing losses and also increases phosphorus content.

Partially rotten farmyard manure has to be applied three to four weeks before sowing while well rotten manure can be applied immediately before sowing. Generally 10 to 20 t/ha is applied, but more than 20 t/ha is applied to fodder grasses and vegetables. In such cases farmyard manure should be applied at least 15 days in advance to avoid immobilisation of nitrogen. The existing practice of leaving manure in small heaps scattered in the field for a very long period leads to loss of nutrients. These losses can be reduced by spreading the manure and incorporating by ploughing immediately after application.

Figure: Farm yard manure

Vegetable crops like potato, tomato, sweet-potato, carrot, raddish, onion etc., respond well to the farmyard manure. The other responsive crops are sugarcane, rice, napier grass and orchard crops like oranges, banana, mango and plantation crop like coconut.

The entire amount of nutrients present in farmyard manure is not available immediately. About 30 per cent of nitrogen, 60 to 70 per cent of phosphorus and 70 per cent of potassium are available to the first crop.

Sheep and Goat Manure

The droppings of sheep and goats contain higher nutrients than farmyard manure and compost. On an average, the manure contains 3 per cent N, 1 per cent P_2O_5 and 2 per cent K_2O. It is applied to the field in two ways. The sweeping of sheep or goat sheds are placed in pits for decomposition and it is applied later to the field. The nutrients present in the urine are *wasted* in this method. The second method is sheep penning, wherein sheep and goats are kept overnight in the field and urine and fecal matter added to the soil is incorporated to a shallow depth by working blade harrow or cultivator or cultivator.

Poultry Manure

The excreta of birds ferment very quickly. If left exposed, 50 percent of its nitrogen is lost within 30 days. Poultry manure contains higher nitrogen and phosphorus compared to other bulky organic manures. The average nutrient content is 3.03 per cent N; 2.63 per cent P_2O_5 and 1.4 per cent K_2O.

Concentrated Organic Manures

Concentrated organic manures have higher nutrient content than bulky organic manure. The important concentrated organic manures are oilcakes, blood meal, fish manure etc. These are also known as organic nitrogen fertilizer. Before their organic nitrogen is used by the crops, it is converted through bacterial action into readily usable ammoniacal nitrogen and nitrate nitrogen. These organic fertilizers are, therefore, relatively slow acting, but they supply available nitrogen for a longer period.

Oil Cakes

After oil is extracted from oilseeds, the remaining solid portion is dried as cake which can, be used as manure. The oil cakes are of two types:
1. Edible oil cakes which can be safely fed to livestock; e.g.: Groundnut cake, Coconut cake etc., and
2. Non edible oil cakes which are not fit for feeding livestock; e.g.: Castor cake, Neem cake, Mahua cake etc.,

Both edible and non-edible oil cakes can be used as manures. However, edible oil cakes are fed to cattle and non-edible oil cakes are used as manures especially for horticultural crops. Nutrients present in oil cakes, after mineralisation, are made available to crops 7 to 10 days after application. Oilcakes need to be well powdered before application for even distribution and quicker decomposition.

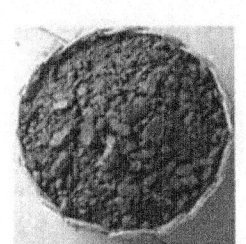

Jatropha oil cakes　　Pongamia oil cakes　　Cottonseed oil cakes

The average nutrient content of different oil-cakes is presented in the following table.

Average Nutrient Content of Oil Cakes

Oil-cakes	Nutrient content (%)		
	N	P2O5	K2O
Non edible oil-cakes			
Castor cake	4.3	1.8	1.3
Cotton seed cake (undecorticated)	3.9	1.8	1.6
Karanj cake	3.9	0.9	1.2
Mahua cake	2.5	0.8	1.2
Safflower cake (undecorticated)	4.9	1.4	1.2
Edible oil-cakes			
Coconut cake	3.0	1.9	1.8
Cotton seed cake (decorticated)	6.4	2.9	2.2
Groundnut cake	7.3	1.5	1.3
Linseed cake	4.9	1.4	1.3
Niger cake	4.7	1.8	1.3
Rape seed cake	5.2	1.8	1.2
Safflower cake (decorticated)	7.9	2.2	1.9
Sesamum cake	6.2	2.0	1.2

Other Concentrated Organic Manures

Blood meal when dried and powdered can be used as manure. The meat of dead animals is dried and converted into meat meal which is a good source of nitrogen. Average nutrient content of animal based concentrated organic manures is given as follows.

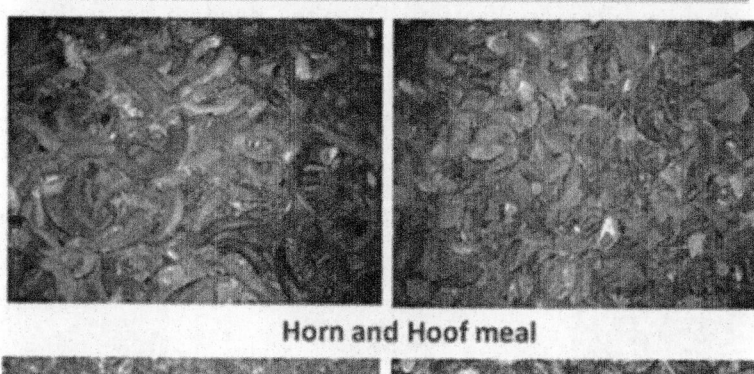

Horn and Hoof meal

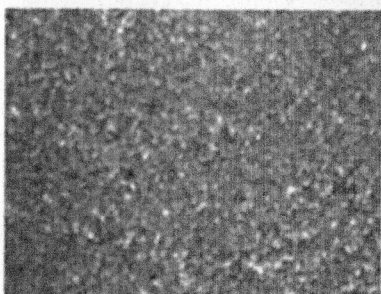

Raw bone meal

Crushed bone meal

Average Nutrient Content of Animal Based Concentrated Organic Manures

Organic manures	Nutrient content (%)		
	N	P2O5	K2O
Blood meal	10 - 12	1 - 2	1.0
Meat meal	10.5	2.5	0.5
Fish meal	4 - 10	3 - 9	0.3 - 1.5
Horn and Hoof meal	13	-	-
Raw bone meal	3 - 4	20 - 25	-
Steamed bone meal	1 - 2	25 - 30	-

What is Organic Farming?

Organic farming system in India is not new and is being followed from ancient time. It is a method of farming system which primarily aimed at cultivating the land and raising crops in such a way, as to keep the soil alive and in good health by use of organic wastes (crop, animal and farm wastes, aquatic wastes) and other biological materials along with beneficial microbes (biofertilizers) to release nutrients to crops for increased sustainable production in an eco friendly pollution free environment. As per the definition of the United States Department of Agriculture (USDA) study team on organic farming "organic farming is a system which avoids or largely excludes the use of synthetic inputs (such as fertilizers, pesticides, hormones, feed additives etc) and to the maximum extent feasible rely upon crop rotations, crop residues, animal manures, off-farm organic waste, mineral grade rock additives and biological system of nutrient mobilisation and plant protection". FAO suggested that "Organic agriculture is a unique production management system which promotes and enhances agro-ecosystem health, including biodiversity, biological cycles and soil biological activity, and this is accomplished by using on-farm agronomic, biological and mechanical methods in exclusion of all synthetic off-farm inputs".

Need of Organic Farming

With the increase in population our compulsion would be not only to stabilise agricultural production but to increase it further in sustainable manner. The scientists have realised that the 'Green Revolution' with high input use has reached a plateau and is now sustained with diminishing return of falling dividends. Thus, a natural balance needs to be maintained at all cost for existence of life and property. The obvious choice for that would be more relevant in the present era, when these agrochemicals which are produced from fossil fuel and are not renewable and are diminishing in availability. It may also cost heavily on our foreign exchange in future.

The key chararacteristics of organic farming include—

1. Protecting the long term fertility of soils by maintaining organic matter levels, encouraging soil biological activity, and careful mechanical intervention;
2. Providing crop nutrients indirectly using relatively insoluble nutrient sources which are made available to the plant by the action of soil micro-organisms;
3. Nitrogen self-sufficiency through the use of legumes and biological nitrogen fixation, as well as effective recycling of organic materials including crop residues and livestock manures;
4. Weed, disease and pest control relying primarily on crop rotations, natural predators, diversity, organic manuring, resistant varieties and limited (preferably minimal) thermal, biological and chemical intervention;
5. The extensive management of livestock, paying full regard to their evolutionary adaptations, behavioural needs and animal welfare issues with respect to nutrition, housing, health, breeding and rearing;
6. Careful attention to the impact of the farming system on the wider environment and the conservation of wildlife and natural habitats.

Bibliography

Alka Rani Upadhyay: *Aquatic Plants for the Waste Water Treatment*, Daya, Delhi, 2004.

Ashworth S.: *Seed to Seed*, Decorah, Seed Savers Publications, 1991.

Balfour, Lady Eve B. *The Living Soil*. London: Faber and Faber, 1943.

Bennett, Hugh Hammond : *Elements of Soil Conservation*, Biotech Books, Delhi, 2009.

Bhakar S.R. : *Ground Water Hydrology : Theory and Practice*, Agrotech, Delhi, 2009.

Bolton, Malcolm D.: *A Guide to Soil Mechanics*, Universities Press, Delhi, 2003.

Borsodi, Ralph. *Flight from the City: An Experiment in Creative Living on the Land*. New York: Harper and Brothers, 1933.

Bourne, Peter G.: *Water and Sanitation: Economic and Sociological Perspectives* Academic Press. Orlando. 1984.

Boyer , J.S.: *Measuring the Water Status of Plants and Soils*, Academic Press, N.Y., 1995.

Brady, Nyle C. *The Nature and Properties of Soils*, Eighth Edition. New York: Macmillan, 1974.

Bromfield, Louis. *Malibar Farm*. New York: Harper & Brothers, 1947.

Carter, Vernon Gill and Dale, Tom. *Topsoil and Civilization*. Norman: University of Oklahoma Press, 1974..

Chandra, Ramesh and Satish Kumar Singh: *Fundamentals and Management of Soil Quality*, Westville Pub, Delhi, 2009.

Dabholkar, A.R. : *General Plant Breeding*, Concept, Delhi, 2006.

Dar, Ghulam Hassan: *Soil Microbiology and Biochemistry*, New India Publishing Agency, Delhi, 2010.

Darwin, Charles R.: *The Formation of Vegetable Mould through the Action of Worms with Observations on their Habits*. London: John Murray & Co., 1981.

Das, Braja M. : *Advanced Soil Mechanics*, Taylor and Francis, Delhi, 2010.

Devi, C.R. Sudharmai : *Analytical Procedures in Soil Science and Agricultural Chemistry*, Agrotech, Delhi, 2004.

Dubey, Sarvesh Kumar and Asha Arora: *A Practical Book on Soil, Plant, Water and Fertilizer Analysis*, S.R. Scientific, Delhi, 2010.

Ferentinos L.: *Proceeding of the Sustainable Taro Culture for the Pacific Conference*, Honolulu, HITAHR, 1993.

Foth, Henry D. *Fundamentals of Soil Science*, Eighth Edition. New York: John Wylie & Sons, 1990.

Golueke, Clarence G., Ph.D. *Composting: A Study of the Process and its Principles*. Emmaus: Rodale Press, 1972.

Gupta, O.P. : *Water in Relation to Soils and Plants : With Special Reference to Agriculture*, Agrobios, Delhi, 2002.

Gupta, P.K. : *A Handbook of Soil Fertilizer and Manure*, Agrobios, Delhi, 2011.

Gustafson, A.F. : *Conservation of the Soil*, Biotech Books, Delhi, 2011.

Herminie Broedel Kitchen: *Soils and Crops : Diagnostic Techniques*, Satish Serial Publishing, Allahabad, 2004.

Hopkins, Cyril G. *Soil Fertility and Permanent Agriculture*. Boston: Ginn and Company, 1910.

Jackson, E. : *Crop Management and Soil Conservation*, Biotech Books, Delhi, 2011.

Jana B L : *Water Harvesting and Watershed Management*, Agrotech, Delhi, 2008.

Jenny, Hans. *Factors of Soil Formation: a System of Quantitative Pedology*. New York: McGraw Hill, 1941.

Jha Timir Baran and Ghosh Biswajit : *Plant Tissue Culture : Basic and Applied*, Universities Press, Delhi, 2005.

Jhonson, Charlotte : *Biology of Soil Science*, Oxford Book Company, Delhi, 2009.

Kataria, T N : *Plant and Crop Physiology*, Pearl Books, Delhi, 2008.

Kevan, D. Keith. *Soil Animals: Methods and Application*. London: H. F. & G. Witherby Ltd., 1962.

Krasilnikov, N A.: *Soil Microorganisms and Higher Plants*. Y.A. Halperin. Jerusalem: Israel Program for Scientific Translations, 1961.

Kumar, N. : *Breeding of Horticultural Crops : Principles and Practices*, New India Pub, Delhi, 2006.

Kumar, Pushpam : *Economics of Soil Erosion : Issues and Imperatives from India*, Concept, Delhi, 2004.

Macself, R : *Soils and Fertilizers*, Satish Serial Pub, Delhi, 2005.

Majumdar, S.P. and R.A. Singh: *Analysis of Soil Physical Properties*, Agrobios, Delhi, 2008.

Malcolm D. Bolton: *A Guide to Soil Mechanics*, Universities Press, Delhi, 2003.

Maliwal, G.L. and K.P. Patel: *Heavy Metals in Soils and Plants*, Agrotech Pub, Delhi, 2011.

Mathur, S K B B S Kapoor: *Emerging Trends In Soil Management*, Madhu Publications, Delhi, 2007.

Nagamani; A ; I K Kunwar and C Manoharachary: *Handbook of Soil Fungi*, I K International, Delhi, 2006.

Naik, M K and G S Devika Rani: *Advances in Soil Borne Plant Diseases*, New India Pub, Delhi, 2008.

Narwal, S.S. : *Allelopathy In Soil Sickness*, Scientific, Delhi, 2006.

Parnes, Robert. *Organic and Inorganic Fertilizers*. Mt. Vernon, Maine: Woods End Agricultural Institute, 1986.

Patel, Sheelwant: *Indian Forests : Soil Water and Bio-Environment Conservation*, Pointer, Delhi, 2005.

Pfeiffer, E.E.: *Biodynamic Farming and Gardening*. Spring Valley, New York: Anthroposophic Press, 1938.

Prasad, T.V.S. : *A Textbook of Soil Microbiology*, Dominant Pub, Delhi, 2011.

Price, Weston A. *Nutrition and Physical Degeneration*. La Mesa, California: Price-Pottenger Nutrition Foundation, reprinted 1970. (1939)

Rao, A.V. Narasimha: *Fundamentals of Soil Mechanics*, Laxmi Publications, Delhi, 2010.

Rob Jenkins and C.K. Jain: *Advances in Soil-Borne Plant Diseases*, Oxford Book Company, Delhi, 2010.

Russell, E.J. : *Handbook on Soils and Manures*, Discovery, 1999.

Russell, Sir E. John. *Soil Conditions and Plant Growth*. New York: Longmans, Green & Co., 1950.

Sambamurty, A V S S : *Textbook of Plant Pathology*, I K Pub, Delhi, 2006.

Samuel W. Johnson: *Crops Feed from Air and Soil*, Reprint Pub, Delhi, 2005.

Schuphan, Werner. *Nutritional Values in Crops and Plants*. London: Faber and Faber, 1965.

Sharieff, Afzal : *Fundamentals of Soil Geography*, Sarup Book Pub, Delhi, 2010.

Sharma, Premjit: *Applied Soil Ecology*, Gene Tech Books, Delhi, 2007.

Singh, S K : *Biotechnology, Plant Propagation and Plant Breeding*, Campus Books, Delhi, 2008.

Smith, J. Russell. *Tree Crops: A Permanent Agriculture.* New York: Harcourt, Brace and Company, 1929.

Somani, L L and P C Kanthaliya: *Soils and Fertilisers at a Glance*, Agrotech, Delhi, 2004.

Thakore, Sanat: *Soil Microbiology*, Dominant Pub, Delhi, 2008.

Turner, Frank Newman. *Fertility, Pastures and Cover Crops Based on Nature's Own Balanced Organic Pasture Feeds.* reprinted from: Faber and Faber, 1955.

Whealy K.: *The Garden Seed Inventory*, Decorah, Seed Saver Publications, 1988.

Wilde, S.A. : *Forest Soils and Forest Growth*, Periodicals, Delhi, 1991.

Index

A

Acidic Soil, 26, 47, 48, 52, 53, 54, 65, 67, 68, 81, 87, 134, 135, 136, 248.
Acidic Soil Plants, 54, 134, 135, 136.

B

Bacteria, 3, 5, 11, 15, 20, 33, 50, 56, 86, 116, 134, 148, 163, 165, 166, 167, 168, 179, 186, 187, 188, 189, 199, 211, 230, 232, 233, 236, 237, 238, 247.
Bulk Density, 23, 43, 220.

C

Capillary Fringe, 17.
Carbon Cycle, 170.
Carbon Sequestration, 198.
Carbon Stock, 235.
Conserve Soil, 91, 92, 93, 111, 113, 116, 117, 118, 123, 155.
Cultivated Soils, 153, 214.

F

Farming System, 256, 258.
Field Capacity, 41, 221.
Food Production, 8, 189, 230.

G

Garden Soil Preparation, 79, 80, 81.
Global Warming, 15, 121, 163.

H

Humus Soil, 36, 37, 38.

I

Igneous Rocks, 133, 139.

L

Laboratory Tests, 249.
Land Pollution, 70, 87, 88, 89, 90, 91.

M

Mass Movement, 120.
Moisture Analysis, 141, 142, 143.
Moisture Detection, 141, 142, 143.
Moisture Metre, 41, 141, 146, 147.

N

Natural Gas, 163.

O

Organic Carbon, 149, 164, 170, 171, 172, 173, 178, 179, 181, 182, 183, 234.
Organic Farming, 112, 184, 185, 186, 187, 189, 190, 191, 192, 193, 195, 197, 198, 199, 200, 201, 256, 257, 258.
Organic Geochemistry, 170.

P

Parent Rocks, 4.
Performance, 62, 86, 172, 178, 179, 180, 193.

Physical Properties, 85, 154, 155, 221, 246, 249, 251, 252.
Plant Nutrition, 201, 204.
Plant Sensitivity, 250.
Potting Soil Recipe, 81, 82, 83, 84.

S

Sedimentary Rocks, 18, 133, 139, 140.
Soil Amendments, 49, 50, 69, 81, 85, 86, 187, 249, 251, 258.
Soil Biodiversity, 230, 231, 232, 234.
Soil Classification, 10, 27, 28, 30.
Soil Compaction, 204, 219, 220, 221, 222, 223.
Soil Conservation, 91, 92, 93, 105, 110, 111, 112, 113, 114, 115, 116, 121, 123, 218.
Soil Contamination, 16, 90, 110, 114, 229.
Soil Electrical Conductivity Variability, 42.
Soil Erosion, 17, 39, 40, 42, 70, 77, 90, 92, 93, 96, 97, 98, 99, 100, 101, 102, 103, 104, 105, 107, 108, 109, 110, 111, 112, 114, 115, 117, 118, 119, 120, 121, 122, 123, 124, 145, 157, 191, 197, 214, 235.
Soil Food Web, 233.
Soil Horizons, 1, 4, 6, 9, 15, 25, 30, 55, 77.
Soil Management, 16, 39, 40, 88, 90, 104, 111, 198, 208, 218.
Soil Moisture, 6, 12, 40, 41, 42, 43, 45, 46, 142, 143, 158, 221.
Soil Organic Matter, 147, 148, 151, 152, 153, 154, 155, 156, 157, 158, 159, 160, 187, 193, 201, 204, 206, 208, 209, 211, 212, 213, 214, 216, 217, 218, 236, 238, 242, 244, 247.
Soil Pollution, 38, 39, 40, 69, 70, 88, 90, 102, 243.
Soil Salination, 17.
Soil Salinisation, 218.
Soil Salinity, 17, 42, 115, 249.
Soil Samples, 45, 57, 58, 71, 72, 74, 75, 76, 130, 131.
Soil Sealing, 228, 229.
Soil Susceptibility, 222, 223.
Soil Taxonomy, 10, 57.
Soil Testing, 47, 49, 51, 57, 65, 66, 68, 129, 130, 131.
Soil Texture, 9, 13, 25, 49, 236, 249.
Soil Thematic Strategy, 227.

T

Trace Elements, 61, 245.

W

Watertable Control, 17.
Wind Erosion, 93, 94, 95, 96, 100, 102, 103, 106, 107, 108, 119, 120, 121, 122.

◻◻◻